D0011500

More praise for *Loneliness*

"Cacioppo . . . is part of the school of evolutionary psychologists . . . that believes our species wouldn't have survived without a cooperative instinct. . . . [*Loneliness*] argues that loneliness, like hunger, is an alarm signal that . . . [is] nature's way of telling us to rejoin the group or pay the price."
—Jennifer Senior, *New York*

"A solid scientific look at the physical and emotional impact of loneliness."
—*Publishers Weekly*

"Cacioppo has come to the conclusion that, by compelling us to seek out our fellow humans, loneliness has played a central role in the development of society."
—Stephen Pincock, *Financial Times Magazine*

"University of Chicago professor Cacioppo . . . [gives] us a whole new view of the dangers of loneliness."
—*Library Journal*

"In carefully outlining the science behind their logic using (mostly) lay language, the authors allow the general public to appreciate the complexities of human behavior while at the same time demonstrating the rigors of scientific investigation."
—Brent A. Mattingly, *Contemporary Psychology: American Psychological Association Review of Books*

"Based on years of research, this magnificent exposé discusses the loneliness many people feel, advising them to reach out to others. Our species naturally reciprocates social gestures."
—Frans de Waal, author of *Our Inner Ape*

"I never imagined that one book could explain so much about human nature. And yet this scientific exploration does not diminish us. Instead, it exalts our simple humanity. *Loneliness* is a beautiful message of human connection and a beautiful book."

—Sidney Poitier, Academy Award–winning actor and author of *The Measure of a Man*

"After reading this book you'll never want to be lonely again—nor will you have to be." —Mihály Csíkszentmihályi, author of *Flow*

"*Loneliness* . . . sounds a wake-up call for those of us walking around in a state of isolation—and we are plenty."

—Heidi Stevens, *Chicago Tribune*

"*Loneliness* presents a scientific look at the impact of loneliness and shows that we are far more intertwined and interdependent than our culture has allowed us to acknowledge. Ultimately, the book demonstrates the irrationality of our culture's intense focus on competition and individualism at the expense of family and community."

—SirReadaLot.org

"Introducing relevant evidence derived from closely controlled university experiments, accompanied by anthropological field observations and animal studies, the authors elucidate the underpinnings of human nature and behavior. . . . A superb complement to John Bowlby's *Loss: Sadness and Depression*. . . . Highly recommended."

—Lynne F. Maxwell, *Library Journal*

"Just as hunger prevents us from starving and pain causes us to retreat from physical danger, the authors help us see that loneliness is a symptom of our basic need to connect. . . . This fascinating, complex, and yet highly accessible exploration reminds us that humans are inherently social creatures and that no child or adult can develop properly in the absence of strong social bonds."

—Melinda Blau, coauthor of *Secrets of the Baby Whisperer*, *Secrets of the Baby Whisperer for Toddlers*, and *The Baby Whisperer Solves All Your Problems*

"John T. Cacioppo . . . is one of the founders of the new, interdisciplinary field of neuroscience which has used brain scans to examine the ways in which social isolation impacts our bodies and behavior. Along with science writer William Patrick, he presents a fascinating assessment of loneliness and the need for social connections."

—Frederic and Mary Ann Brussat, *Spirituality & Practice*

"*Loneliness* . . . builds a compelling case for social connection as a basic human drive (and to the importance of being tuned in to when that drive's not being satisfied)." —Susan Pinker, *Globe and Mail*

"Messrs. Cacioppo and Patrick [argue] . . . that a concerted attack on loneliness would improve public health as well as individual happiness." —Andrew Stark, *Wall Street Journal*

"Anyone with an interest in evolutionary psychology and neuroscience will find [John T. Cacioppo and William Patrick's] exhaustive research fascinating. . . . The authors use humor and give suggestions on how to be less lonely."

—Joyce Boaz, Gift From Within

SELECTED WORKS COAUTHORED BY JOHN CACIOPPO

Handbook of Neuroscience for the Behavioral Sciences
(with Gary G. Bernston)

Social Neuroscience: People Thinking about Thinking People
(with Penny S. Visser and Cynthia L. Pickett)

Essays in Social Neuroscience (with Gary G. Bernston)

Emotional Contagion (with Richard L. Rapson)

Attitudes and Persuasion: Classic and Contemporary Approaches
(with Richard E. Petty)

loneliness

HUMAN NATURE AND THE NEED FOR SOCIAL CONNECTION

John T. Cacioppo

AND

William Patrick

W. W. Norton & Company
New York London

Copyright © 2008 by John T. Cacioppo and William Patrick
Drawings copyright © 2008 by Alan Witschonke Illustration

All rights reserved
Printed in the United States of America
First published as a Norton paperback 2009

For information about permission to reproduce selections from this book, write to
Permissions, W. W. Norton & Company, Inc., 500 Fifth Avenue, New York, NY 10110

For information about special discounts for bulk purchases, please contact W. W. Norton
Special Sales at specialsales@wwnorton.com or 800–233–4830

Manufacturing by LSC Harrisonburg
Book design by Wesley Gott
Production manager: Julia Druskin

Library of Congress Cataloging-in-Publication Data
Cacioppo, John T.
Loneliness : human nature and the need for social connection / John T. Cacioppo and
William Patrick.
p. cm.
Includes bibliographical references and index.
ISBN 978–0–393–06170–3 (hardcover)
1. Loneliness. 2. Loneliness—Physiological aspects. 3. Neuropsychology. I. Patrick,
William, date. II. Title.
BF575.L7C23 2008
155.9'2—dc22 2008015099

ISBN 978-0-393-33528-6 pbk.

W. W. Norton & Company, Inc., 500 Fifth Avenue, New York, N.Y. 10110
www.wwnorton.com

W. W. Norton & Company Ltd.
15 Carlisle Street, London W1D 3BS

12 13 14 15 16 17 18 19 20

For Wendi and Carolyn

contents

acknowledgments

The writing of this book was a collaboration involving two authors—an invaluable form of social connection—yet only one was a participant in the more than twenty years of scientific research that is the foundation of the story. Thus, for the sake of convenience and clarity, we chose to write in the first-person singular with John Cacioppo as the narrative voice. We employ that convention in these acknowledgments as well.

But even the research that "I," John Cacioppo, conducted was never a solo effort. That research on social connection began in the early 1990s at Ohio State University, where I taught. We (my scientific colleagues and I) began with the simple question of what are the effects of human association. To address this question, we first conducted experiments in which individuals were randomly assigned to be alone or with others of various kinds (e.g., friends, strangers) while performing a task. We quickly surmised that it was an individual's perceptions of the social situation that mattered most. We moved from an interest in social support to an interest in perceived social isolation—loneliness—as a model system for studying the role of the social world in human biology and behavior. Doing so changed how we conceived the human mind as well.

The dominant metaphor for the scientific study of the human mind during the latter half of the twentieth century has been the computer—a solitary device with massive information processing capacities. Our studies of loneliness left us unsatisfied with this metaphor. Computers today are massively interconnected devices with capacities that extend far beyond the resident hardware and software of a solitary computer. It became apparent to us that the telereceptors (e.g., eyes, ears) of the human brain have provided wireless broadband interconnectivity to humans for millennia. Just as computers have capacities and processes that are transduced through but extend far beyond the hardware of a single computer, the human brain has evolved to promote social and cultural capacities and processes that are transduced through but extend far beyond a solitary brain. To understand the full capacity of humans, one needs to appreciate not only the memory and computational power of the brain but its capacity for representing, understanding, and connecting with other individuals. That is, one needs to recognize that we have evolved a powerful, meaning-making *social* brain.

The notion that humans are inherently social creatures is no longer contestable, but what precisely this means for lives and societies is not fully appreciated either. Governments worldwide rely on economic advisors while publicly mocking scientific studies of social relationships. In an issue of the popular science magazine *Scientific American*, the editors observed that "whenever we run articles on social topics, some readers protest that we should stick to 'real' science." The editors went on to say:

> Ironically, we seldom hear these complaints from working physical or biological scientists. They are the first to point out that the natural universe, for all its complexity, is easier to understand than the human being. If social science seems mushy, it is largely because the subject matter is so difficult, not because humans are somehow unworthy of scientific inquiry. ("The Peculiar Institution," April 30, 2002, p. 8)

The fact that loneliness is unpleasant is obvious. In Genesis, Adam and Eve's punishment for disobeying God was their exile from Eden. In Ovid's *Metamorphoses*, Zeus decided to destroy the men of the Bronze Age by flooding Hellas. Deucalion survived by constructing a chest and, with Pyrrha, drifted to Parnassus. Deucalion realized that however difficult or impossible it is to live with others, even more difficult and more impossible is it to live without them, in complete loneliness. When Zeus granted him to choose what he wished, he chose to create others. However, the notion that loneliness plays an important function for humans, just as do physical pain, hunger, and thirst, and that understanding this function and its effects on social cognition holds some of the secrets to healthier, wealthier, happier lives is not so apparent.

Perhaps fittingly, this book on the science of social connection reflects the contributions of many brilliant and wonderful colleagues, friends, students, and staff. Our scientific studies of the causes, nature, and consequences of loneliness and social connection have ranged across disciplinary, institutional, and international boundaries. The research has included genetic, immunologic, endocrinologic, autonomic, brain imaging, behavioral, cognitive, emotional, personological, social psychological, demographic, and sociological analyses. The range of studies we sought to conduct exceeded my expertise, so scientists from various disciplines have contributed their time, expertise, and insights. These scientific collaborations provided synergies that transformed the research we were able to conduct and amplified the scientific story that unfolded before us.

We (William Patrick and I) wish to thank all the individuals who volunteered to participate in this research over the past two decades. Without their participation and assistance, none of this would have been possible. This book describes the stories of a number of the individuals we studied or interviewed to give a face to our findings. Bill and I have changed the names of and various irrelevant details about these individuals to ensure their true identity is protected. In the case of Katie Bishop, an individual whose case we return to repeatedly in the book, we have used a composite character. This was done to protect the confidentiality of the individuals who participated in our studies.

Among those who are owed special recognition and thanks are Louise Hawkley (University of Chicago), a close scientific collaborator on all aspects of this research for more than a decade, and Gary Berntson (Ohio State University), a close collaborator for the past two decades. In addition, Jan Kiecolt-Glaser (Ohio State University Medical School), William Malarkey (Ohio State University Medical School), Ron Glaser (Ohio State University Medical School), Michael Browne (Ohio State University), Robert MacCallum (University of North Carolina), Phil Marucha (University of Illinois Chicago), Bert Uchino (University of Utah), John Ernst (Illinois Wesleyan University), Mary Burleson (Arizona State University), Tiffany Ito (University of Colorado), Mary Snydersmith (Ohio State University), Kirsten Poehlmann (University of California San Diego), Ray Kowalewski (Microsoft Corporation), David Lozano (Mindware Corp.), Alisa Paulsen (Ohio State University), and Dan Litvack (Ohio State University) played critical roles in the early stages of our program of research.

In the mid-1990s, I joined the John D. and Catherine T. MacArthur Foundation Network on Mind-Body Integration, directed by Robert Rose. Bob further fueled our interest in the causes, nature, and consequences of loneliness, and the rest of the MacArthur Network members were generous with their time and expertise. We thank Bob and the other Network members, David Spiegel (Stanford University), Esther Sternberg (National Institutes of Health), William Lovallo (University of Oklahoma Health Sciences Center), Kenneth Hugdahl (University of Bergen), Eve Van Cauter (University of Chicago), J. Allan Hobson (Harvard University), John Sheridan (Ohio State University), Steve Kosslyn (Harvard University), Martha McClintock (University of Chicago), Anne Harrington (Harvard University), and Richard Davidson (University of Wisconsin), for their suggestions, assistance, and support.

When I moved to the University of Chicago in 1999, we were joined by additional gifted and generous scholars from the social and biological sciences, including Linda Waite (University of Chicago), Ronald Thisted (University of Chicago), M. E. Hughes (Johns Hopkins University), Christopher Masi (University of Chicago), Steve Cole (University of California, Los Angeles),

Thomas McDade (Northwestern University), Emma Adam (Northwestern University), Ariel Kalil (University of Chicago), Brigitte Kudielka (University of Trier), Howard Nusbaum (University of Chicago), W. Clark Gilpin (University of Chicago), Dorret Boomsma (Free University Amsterdam), Penny Visser (University of Chicago), Jean Decety (University of Chicago), Tanya Luhrmann (Stanford University), Farr Curlin (University of Chicago), Gün Semin (Utrecht University), Kellie Brown (Medical College of Wisconsin), Ming Wen (University of Utah), L. Elizabeth Crawford (Richmond University), Jarett Berry (Northwestern University Medical School), Kristopher Preacher (University of Kansas), Nick Epley (University of Chicago), Adam Waytz (University of Chicago), Steve Small (University of Chicago), Kathryn Tanner (University of Chicago), Omar McRoberts (University of Chicago), Roberto Lang (University of Chicago), Roy Weiss (University of Chicago), George Monteleone (University of Chicago), Jos Bosch (University of Birmingham), Chris Engeland (University of Illinois at Chicago), Phil Schumm (University of Chicago), Edith Rickett (University of Chicago), Diana Greene (University of Chicago), Kathleen Ziol-Guest (Harvard University), Catherine Norris (Dartmouth College), Matthew Christian (University of Chicago), Ken Olliff (University of Chicago), Jeffrey Darragh (University of Chicago), and Barnaby Marsh (Oxford University). We owe thanks, too, to the many other staff and students who have worked so diligently with us over the years.

Scientific research of the kind described in this book is costly. We therefore are grateful for the research support over the years from the National Institute on Aging Grant No. PO1 AG18911, the National Science Foundation Grant No. BCS-0086314, the National Institute of Mental Health Grant No. P50 MH72850, the John D. and Catherine T. MacArthur Foundation, and the John Templeton Foundation. The views contained in this book are those of the authors, of course, but without funding for our basic scientific research we would have considerably less to say on the topic of loneliness and social connection.

Finally, we thank Lisa Adams, who not only convinced us to

undertake this book but made it possible for us to do so, and Maria Guarnaschelli of Norton, who has been everything and more that one could hope to find in an editor. Camille Smith, who as a manuscript editor over the years has made hundreds of writers and academics look smarter than they are, has done the same for us. Thank you, Camille. And most important, we thank our families and our spouses, Wendi and Carolyn, for teaching us the value and power of social connection.

This book is about life, loneliness, and the power of social connection. It is perhaps ironic that while working on this book we both lost our mothers. Each lived a full and happy life, and each passed peacefully surrounded by family and friends. And although expected, their deaths set off a tsunami of feelings and emotions as we dealt with the loss of our first connection to another person. Dealing with their loss made us even more grateful than before to our family and friends for their kindness and generosity, more convinced of the fundamental importance of social connections, and more empathetic to those who live in the shroud of perceived social isolation. We dedicate this book to the memory of our mothers, Mary Katherine Cacioppo and Bernice Turner.

John T. Cacioppo and William Patrick

If you want to go fast, go alone. If you want to go far, go together.

—*African Proverb*

PART ONE

the lonely heart

I am fifty-six and have been divorced for years. When I was still with my husband and told someone I was lonely they responded with "but you're married." I have learned the difference between being alone and lonely. In a crowd, at work, even in a family setting, I always feel lonely. It can be overwhelming at times, a physical sensation. My doctors have called it depression, but there is a difference. I read once, you are born alone and you die alone. But what about all the years in between? Can you really belong to someone else? Can you ever resolve the inner feeling of being alone? Shopping won't do it. Eating won't do it. Random sex doesn't make it go away. If and when you find any answers, please write back and tell me.

—*Letter from a woman who read about our research in a magazine*

CHAPTER ONE

lonely in a social world

Katie Bishop grew up surrounded by aunts and uncles, grandparents and cousins, in a small community that was nothing if not closely knit. Between family events, church events, sports, and music, her entire childhood was spent among the same friendly people. Truth be told, she could hardly wait to get away. Despite all the togetherness, she always felt a little out of it, and by the time she graduated from high school she was ready for a change. She did not have enough money to go away for college, so for the next four years she lived at home and commuted. But the moment she had her degree, she moved about as far away as she could to take a job in the software industry.

Katie's new career required her to spend weeks at a time hopping around from city to city. She still talked to her mother and her sister once or twice a week, but now the contact was mediated through her Blackberry, her laptop, or the phone in her kitchen. After six months of this very different routine she realized that she was not sleeping well. In fact, her whole body seemed to be off. If a cold or flu bug was anywhere in her vicinity, she would catch it. When she wasn't traveling or working long hours, or taking yoga classes to try to deal with the back and neck pain from traveling and working those long hours, she spent a great deal of time in front of the TV, eating ice cream straight from the carton.

Six months into her new, independent life, Katie Bishop was fifteen pounds heavier and truly miserable. She didn't just feel fat, she felt ugly. And after an unpleasant run-in at the home office and a spat with one of her neighbors, she was even beginning to wonder if she would ever be socially acceptable outside the little town that had made her feel so trapped.

It doesn't take a degree in psychology to figure out that Katie Bishop was lonely. But Katie's loneliness was more than just the mild heartache that fuels pop songs and Miss Lonelyhearts columns. Katie was dealing with a serious problem that has deep roots in her biology as well as her social environment. It began with a genetic predisposition that set her standards for social connection very high, although we might also express it as a high sensitivity to feeling the absence of connection. There is certainly nothing wrong with having high standards, but this physiological need, set against an environment that failed to satisfy that need, was beginning to distort her perceptions and her behavior. It was also setting in motion a series of cellular events that might seriously compromise her health.

While growing up in that tightly knit community, Katie never gave much thought to social connection one way or the other. As a kid she could be cranky at times, a little difficult, and sometimes her parents assumed she was depressed. One of her English teachers, assigning it almost as a badge of honor, described Katie as "alienated." A more accurate description would have been that, even as a kid, even while surrounded by family and other friendly people, Katie had always felt a subjective sense of social isolation. By Katie's internal measure, the connections in her world seemed somehow fragile and distanced. She could not consciously articulate what was bothering her, but as soon as she could, she opted for a dramatic change of scene. She thought that being entirely on her own would be just what she needed. In fact, what she needed was not *less* social connection, but connection that felt more meaningful—a level of connection that matched her genetically biased predisposition.

Almost everyone feels the pangs of loneliness at certain moments. It can be brief and superficial—being the last one chosen for a team on the playground—or it can be acute and severe—suffering the death of a spouse or a dear friend. Transient loneliness is so common, in fact, that we simply accept it as a part of life. Humans are, after all, inherently social beings. When people are asked what pleasures contribute most to happiness, the overwhelming majority rate love, intimacy, and social affiliation above wealth or fame, even above physical health.[1] Given the importance of social connection to our species, then, it is all the more troubling that, at any given time, roughly twenty percent of individuals—that would be sixty million people in the U.S. alone—feel sufficiently isolated for it to be a major source of unhappiness in their lives.[2]

This finding becomes even more compelling when we consider that social isolation has an impact on health comparable to the effect of high blood pressure, lack of exercise, obesity, or smoking.[3] Our research in the past decade or so demonstrates that the culprit behind these dire statistics is not usually being literally alone, but the subjective *experience* known as loneliness. Whether you are at home with your family, working in an office crowded with bright and attractive young people, touring Disneyland, or sitting alone in a fleabag hotel on the wrong side of town, chronic *feelings* of isolation can drive a cascade of physiological events that actually accelerates the aging process. Loneliness not only alters behavior but shows up in measurements of stress hormones, immune function, and cardiovascular function. Over time, these changes in physiology are compounded in ways that may be hastening millions of people to an early grave.

To measure a person's level of loneliness, researchers use a psychological assessment tool called the UCLA Loneliness Scale, a list of twenty questions with no right or wrong answers. It is reproduced here as Figure 1. The questions are not based on information but on very common human feelings. When I refer to people who are lonely or "high in loneliness," I mean those who, regardless of their objective circumstances, score high on this pencil-and-paper test.

If you would like to take the test yourself, I explain how to score it in note 4 on page 271.[4]

*1. How often do you feel that you are "in tune" with the people around you? ____
2. How often do you feel that you lack companionship? ____
3. How often do you feel that there is no one you can turn to? ____
4. How often do you feel alone? ____
*5. How often do you feel part of a group of friends? ____
*6. How often do you feel that you have a lot in common with the people around you? ____
7. How often do you feel that you are no longer close to anyone? ____
8. How often do you feel that your interests and ideas are not shared by those around you? ____
*9. How often do you feel outgoing and friendly? ____
*10. How often do you feel close to people? ____
11. How often do you feel left out? ____
12. How often do you feel that your relationships with others are not meaningful? ____
13. How often do you feel that no one really knows you well? ____
14. How often do you feel isolated from others? ____
*15. How often do you feel you can find companionship when you want it? ____
*16. How often do you feel that there are people who really understand you? ____
17. How often do you feel shy? ____
18. How often do you feel that people are around you but not with you? ____
*19. How often do you feel that there are people you can talk to? ____
*20. How often do you feel that there are people you can turn to? ____

FIGURE 1. The UCLA Loneliness Scale (version 3). From Daniel W. Russell, "UCLA Loneliness Scale (version 3): Reliability, validity, and factor structure," *Journal of Personality Assessment* 66 (1996). Used with permission.

Keep in mind, however, that we can all slip in and out of loneliness. Feeling lonely at any particular moment simply means that you are human. In fact, a sizable portion of this book is devoted to demonstrating that the need for meaningful social connection, and the pain we feel without it, are defining characteristics of our species. Loneliness becomes an issue of serious concern only when it settles in long enough to create a persistent, self-reinforcing loop of negative thoughts, sensations, and behaviors.

Keep in mind, too, that feeling the pain of isolation is not an unalloyed negative. The sensations associated with loneliness evolved because they contributed to our survival as a species. "To be isolated from your band," wrote John Bowlby, the developmental psychologist who pioneered attachment theory, "and, especially when young, to be isolated from your particular caretaker is fraught with the greatest danger. Can we wonder then that each animal is equipped with an instinctive disposition to avoid isolation and to maintain proximity?"[5]

Physical pain protects the individual from physical dangers. Social pain, also known as loneliness, evolved for a similar reason: because it protected the individual from the danger of remaining isolated. Our forebears depended on social bonds for safety and for the successful replication of their genes in the form of offspring who themselves survived long enough to reproduce. Feelings of loneliness told them when those protective bonds were endangered or deficient. In the same way that physical pain serves as a prompt to change behavior—the pain of burning skin tells you to pull your finger away from the frying pan—loneliness developed as a stimulus to get humans to pay more attention to their social connections, and to reach out toward others, to renew frayed or broken bonds. But here was a pain that prompted us to behave in ways that did not always serve our immediate, individual self-interest. Here was a pain that got us outside ourselves, widening our frame of reference beyond the moment.

In English, we have a word for pain and a word for thirst, but no single, specific terms that mean the opposite. We merely reference the absence of these aversive conditions, which makes sense, because their absence is considered part of the normal state. Our research suggests that "not lonely"—there is no better, more specific term for it—is also, like "not thirsty" or "not in pain," very much part of the normal state. Health and well-being for a member of our species requires, among other things, being satisfied and secure in our bonds with other people, a condition of "not being lonely" that, for want of a better word, we call social connection.

And this idea of loneliness as social pain is more than a metaphor. Functional magnetic resonance imaging (fMRI) shows us that the emotional region of the brain that is activated when we experience rejection is, in fact, the same region—the dorsal anterior cingulate—that registers emotional responses to physical pain (see Figure 2).

The discovery that feelings of social rejection (isolation) and reactions to physical pain share the same hardware begins to suggest why, once loneliness becomes chronic, you cannot escape it merely by "coming out of your shell," losing weight, getting a fashion makeover, or meeting Mr. or Ms. Right. The pain of loneliness is a deeply disruptive hurt. The disruption, both physiological and behavioral, can turn an unmet need for connection into a chronic condition, and when it does, changing things for the better requires taking into account the full depth and complexity of the role loneliness plays in our biology and in our evolutionary history. Following Katie Bishop's lead and trying to make ourselves feel better with fatty foods and reruns of *Friends* will only make matters worse.

Connecting the Dots

I have been working for more than thirty years to unravel how our brain and body are intertwined with our social responses. I teach psychology at the University of Chicago, and I direct the Center for Cognitive and Social Neuroscience there. I am also fortunate to be part of a widespread network of partners in this research. These

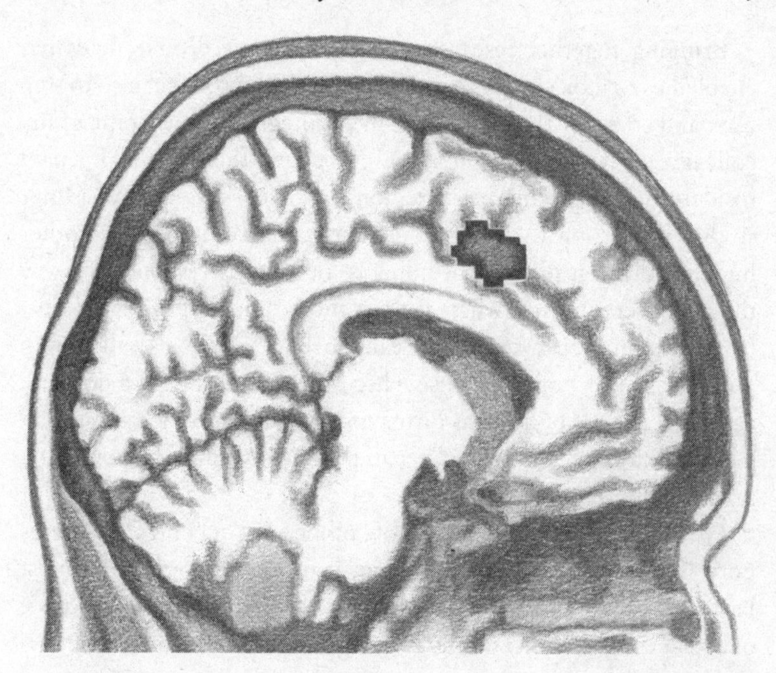

FIGURE 2. The human brain reacting to social pain. The dark rectangular blotch near the top of the brain represents the activation of the dorsal anterior cingulate cortex in response to social rejection. The brain responds similarly to physical pain. Adapted from N. I. Eisenberger, M. Lieberman, and K. D. Williams, "Does rejection hurt? An fMRI study of social exclusion," *Science* 302 (10 October 2003): 290–292.

include present and former colleagues at The University of Chicago and the Ohio State University, as well as a team of psychologists and psychiatrists, sociologists and biostatisticians, cardiologists and endocrinologists, behavioral geneticists and neuroscientists called the MacArthur Mind-Body Network; a similarly diverse team called the MacArthur Aging Society Network; and the Templeton–University of Chicago Research Network, whose members, ranging from neurologists to theologians, from biostatisticians to philosophers, work together to try to understand the links between our physiological responses and our social and even spiritual strivings.

Bringing together researchers from so many diverse fields has enabled us to look closely at each piece of the puzzle, but also to step back and consider the big picture in an integrated way. Some of my colleagues have taken brain scanning beyond the pathway for pain to identify the specific brain regions involved in empathy.[6] Other studies relying on fMRI show us that when we humans see other humans, or even pictures of humans, our brains respond in a way that is different from when we see most other types of objects.[7] (Interestingly, pet owners who really love their animals will show a glimmer of this brain response when shown a picture of a dog or a cat.) And images of humans displaying intense emotion rather than a neutral expression also register in the brain with correspondingly greater intensity.[8]

Given the special importance of "other human beings" as a category reflected in our neural wiring, it makes sense that the most basic rituals of human societies everywhere reflect the importance of social context. For as long as our species has left traces, the evidence suggests that the most emotionally evocative experiences in life have been weddings, births, and deaths—events associated with the beginnings and endings of social bonds. These bonds are the centripetal force that holds life together. The special balm of acceptance that these bonds provide, and the uniquely disturbing pain of rejection when they are denied, is what makes humans so highly attuned to social evaluation. We care deeply what others think of us, and this is why, of the ten most common phobias that cause people to seek treatment, three have to do with social anxiety: fear of speaking in public, fear of crowds, fear of meeting new people.[9]

In trying to understand the tremendous power of social connections and interactions within our own species, some scientists have traced the roots of social impulses all the way back to "avoidance" in octopi and "extroversion" in guppies. Scientists working with social insects find that the connections are so tight that it is easy to think of the hive or the ant hill as a single, extended organism.

Among our fellow mammals, we see social connections that are familiar—wolves teaming up to coordinate the hunt, howling together before and after—and some that are surprising—these same fierce carnivores bringing back meat for packmates who are disabled or who are nursing pups. We see altruistic self-sacrifice in prairie dogs when one individual calls out the first alert when a hawk swoops down, even though this warning action makes it the predator's prime target. And in ape societies, as in every human culture ever studied, we see infractions against the social order being punished by the denial of social connection—the deliberately induced pain known as ostracism. As hominids evolved into humans, and as troops became tribes and cultures became kingdoms, the pain of banishment remained the most severe punishment, short of torture or death, imposed by kings and potentates.[10] It is no accident that even today, in modern correctional institutions, the penalty of last resort is solitary confinement.

The roots of our human impulse for social connection run so deep that feeling isolated can undermine our ability to think clearly, an effect that has a certain poetic justice to it, given the role of social connection in shaping our intelligence. Most neuroscientists now agree that, over a period of tens of thousands of years, it was the need to send and receive, interpret and relay increasingly complex social cues that drove the expansion of, and greater interconnectedness within, the cortical mantle of the human brain. In other words, it was the need to deal with other people that, in large part, made us who and what we are today.[11]

It should not be surprising, then, that the sensory experience of social connection, deeply woven into who we are, helps regulate our physiological and emotional equilibrium. The social environment affects the neural and hormonal signals that govern our behavior, and our behavior, in turn, creates changes in the social environment that affect our neural and hormonal processes. To take an example from a fellow primate, higher levels of testosterone in male rhesus monkeys have been shown to promote sexual behavior; but those same testosterone levels are, in turn, influenced by the availability of receptive females on the social scene nearby.[12] Running is usually an activity that promotes a healthy brain, but in studies conducted with lab rats, running proved less beneficial to the brains of animals housed in social isolation.[13] In humans, loneliness itself has been shown to predict the progression of Alzheimer's disease.[14] And one of our recent studies suggests that loneliness actually has the power to alter DNA transcription in the cells of your immune system.[15]

In these and myriad other ways, feelings of social connection, as well as feelings of disconnection, have an enormous influence on our bodies as well as our behaviors. We all decline physically sooner or later, but loneliness can increase the angle of the downward slope. Conversely, healthy connection can help slow that decline. Once we move into the realm of "high in social well-being"—and this is possible for any of us—we benefit from positive, restorative effects that can help keep us going longer and stronger.

Who Gets Lonely?

No one disputes that being the new kid at school, losing your spouse, or outliving your friends can make meaningful connection more of a challenge. Objective circumstances do matter. Marriage, for instance, can help blunt the sense of feeling alone. Married people are, on average, less lonely than unmarried people, but, then again, marriage is no guarantee. Being miserably lonely inside a marriage has been a literary staple from *Madame Bovary* to *The Sopranos.* And being in a marriage can sometimes limit opportunities for forming other attachments, even platonic ones. Talent, financial success, fame, adoration—none of these offers protection from the subjective experience of being alone. The Sixties icon Janis Joplin, who was as isolated off stage as she was intensely bonded with others while performing, said shortly before her death that she was working on a tune called "I just made love to twenty-five thousand people, but I'm going home alone." Three of the most idolized women of the twentieth century, Judy Garland, Marilyn Monroe, and Diana, Princess of Wales, were famously isolated people. The same was true of Marlon Brando and other legendary leading men.

And yet being alone does not necessarily mean being lonely. In his book *Solitude*, the psychiatrist Anthony Storr explores—in fact recommends—the pleasures of sometimes being by yourself. Think of a naturalist doing research in the rain forest, or a pianist in a marathon practice session, or a bicyclist training in the mountains. Prayer and meditation, as well as scholarship and writing, also involve long stretches of solitude, as do most artistic or scientific endeavors. Needing "time for myself" is one of the great complaints of men and women in today's harried marriages, whether they are multitasking their two careers and family or one spouse is putting in sixty-hour weeks at the office while the other stays home with the kids. In fact, fairly or not, people often judge individuals who are unable to tolerate solitude as being needy or neurotic.

Accordingly, there are no easy-to-assign labels where loneliness is concerned. When a deranged man named Russell Weston Jr. stormed the U.S. Capitol in 1998, his picture appeared on the cover of *Newsweek* under a banner headline: "The Loner." The media applied that same vague judgment to the Unabomber Ted Kaczynski, to President Reagan's assailant John Hinckley, to the Virginia Tech mass murderer Cho Seung-Hui, and to any number of other socially marginalized individuals.

However, our studies of a diverse group of healthy young adults show that everyday folks who feel the pain of isolation very acutely—people who may feel tremendously lonely—have no more in common with the dangerously troubled souls who make headlines than does anyone else. There are extremes within any population, but on average, at least among young adults, those who feel lonely actually spend no more time alone than do those who feel more connected. They are no more or less physically attractive than average, and they do not differ, on average, from the nonlonely in terms of height, weight, age, education, or intelligence. Most important, when we look at the broad continuum (rather than just the extremes) of people who feel lonely, we find that they have the capacity to be just as socially adept as anyone else. Feeling lonely does not mean that we have deficient social skills.[16] Problems arise when feeling lonely makes us less likely to employ the skills we have.

The Problem in Three Parts

The powerful effects of loneliness stem from the interplay of three complex factors that I want to explore with you in depth. These are:

1. *Level of vulnerability to social disconnection.* Each of us inherits from our parents a certain level of need for social inclusion (also expressed as sensitivity to the pain of social exclusion), just as we inherit a certain basic body type and basic level of intelligence. (In each case, the influence of the environment on where that genetic inheritance takes us is also vitally important.) This individual, genetically rooted propensity operates like a thermostat, turning on and off distress signals depending on whether or not our individual need for connection is being met.
2. *Ability to self-regulate the emotions associated with feeling isolated.* Successful self-regulation means being able to cope with challenges while remaining on a fairly even keel—not just outwardly, but deep inside. As loneliness increases and persists, it begins to disrupt some of this ability, a "disregulation" that, at the cellular level, leaves us more vulnerable to various stressors, and also less efficient in carrying out soothing and healing functions such as sleep.
3. *Mental representations and expectations of, as well as reasoning about, others.* Each of us frames our experience through our own perceptions, which makes each of us, to some extent, the architect of our own social world. The sense we make of our interactions with others is called social cognition. When loneliness takes hold, the ways we see ourselves and others, along with the kinds of responses we expect from others, are heavily influenced by both

our feelings of unhappiness and threat and our impaired ability to self-regulate.

Some people love hot sauce—they crave it on everything. For others, a hint of jalapeno sends them gasping for ice water. Human variation in the desire for connection is similarly broad. Some people's personal need for inclusion or sensitivity to exclusion is low enough that they can tolerate moving away from friends or family without too much distress. Others have been shaped by genes and environment to need daily immersion in close social contact in order to feel okay. For those who are more easily distressed, it is the interplay of self-regulation and social cognition that determines what happens next. One person will manage to cope until the next opportunity for connection comes along, while another may enter into a downward spiral of self-defeating, even self-destructive thoughts and behaviors, the kind that encourage cellular responses which, over the long haul, prove dramatically corrosive.

Whatever our own individual sensitivity, our well-being suffers when our particular need for connection has not been met. Because early humans were more likely to survive when they stuck together, evolution reinforced the preference for strong human bonds by selecting genes that support pleasure in company and produce feelings of unease when involuntarily alone. Moreover, and central to the theme of this book, evolution fashioned us not only to feel good when connected, but to feel secure. The vitally important corollary is that evolution shaped us not only to feel bad in isolation, but to feel insecure, as in physically threatened. As we will see, once these feelings arise, social cognition can take the sense of danger and run with it.

The person who starts out with a painful, even frightening sensation of being alone may begin to see dangers everywhere on the social landscape. Filtered through the lens of lonely social cognition, other people may appear more critical, competitive, denigrating, or otherwise unwelcoming. These kinds of interpretations quickly become expectations, as loneliness turns the perfectly normal fear of negative evaluation into a readiness to fend off blows. And then the plot thickens. The fear that can force us into a defensive crouch can also cost us some of our ability to self-regulate. When loneliness is protracted, impaired regulation, combined with distorted social cognition, makes us less likely to acknowledge someone else's perspective. We may become less able to evaluate other people's intentions, which can make us socially awkward, but can also make us vulnerable to manipulation by anyone trying to conceal ulterior motives. At the same time, fear of attack fosters a greater tendency to preemptively blame others. Sometimes this fear makes us lash out. Sometimes it makes us desperate to please, and sometimes it causes us to play the victim.

The sad irony is that these poorly regulated behaviors, prompted by fearful sensations, often elicit the very rejection that we all dread the most. Even more confounding, over time, the feeling of vulnerability that comes with loneliness can make us more likely to be dissatisfied with, and distrustful of, whatever social connections we have. A young bride once took her new husband to task for buying the wrong kind of jelly. The fact that he had gone to the grocery store and stocked the refrigerator earned him no points. "You know I hate grape," she told him. In fact, the subject of jams and jellies had never come up. He thought he was doing something nice to make their new home in a new community more comfortable. But in her mind, he was intentionally disregarding her preferences. Unable to dispel the sense of hurt, she dissolved into a tearful rant. We may reasonably suspect that the real issue for her wasn't jelly, but doubts and fears about the marriage, which generated the feeling of isolation and exposure to threat that we call loneliness.

When we feel isolated, we perceive ourselves as doing all we can on behalf of our relationships, even when all objective evidence indicates otherwise. It is the lonely roommate who throws around snide comments all evening, and then when she meets resistance to the insults says, "You're always criticizing me!" When this leads to an argument, she may be the one who starts to yell, requiring others to raise their voices ever so slightly as they try to reason with her. "Stop yelling at me!" is a not-unlikely response from someone whose social cognition perceives a world that is threatening on all sides, and whose ability to self-regulate has been disrupted by those same perceptions.

The same sort of distortions can affect intimate relationships and persist for years. One partner in a relationship has a higher need for connection than the other currently fulfills—perhaps than the other *can* fulfill. Maybe this other partner is cold and narcissistic, but then again, maybe his or her genes and life experience have simply provided a different (and lower) level of need. The point is not to assign "blame" to one or the other, but to recognize that there is a mismatch. Unfortunately, the partner whose need is unmet may begin to act in ways that the other considers "difficult" or "too demanding" or "needy," which causes him or her to pull away even further, leaving the partner who already feels lonely feeling even more neglected and isolated, which propels the pattern spiraling downward toward greater unhappiness. Seeing this familiar dynamic through the lens of loneliness, and sometimes through the lens of genetically biased—and individually different—levels of need for connection, can allow us to address the problem and the search for solutions at a deeper level.

Just as anyone can feel lonely from time to time, anyone can make a mistake that triggers social anxiety and prompts self-protective thoughts and actions. Certainly school, work, and family life present plenty of moments when it is reasonable to anticipate occasional criticism, attack, or even treachery and betrayal. The key difference is that loneliness causes us to apply these defensive perceptions to situations that are neutral or benign. These negative expectations then have a way of becoming self-fulfilling prophecies.

But even as dismal as this interpersonal dance may appear, the fact that loneliness makes us unwittingly contribute to the choreography is actually a plus. The same social cognition that amplifies the problem also gives us a point of access. The way we frame reality through the filter of our own thoughts is something that, with effort, we can learn to modify. The sense of threat we unconsciously ratchet up is something we can learn to very consciously tone down.

Taking Charge

It has always seemed to me that certain public figures appear perpetually isolated (think Prince Charles), while others appear warm and personally magnetic (think Oprah). In private life, too, some people seem to be natural social connectors, those who bond easily with others and whom everyone enjoys being around. They are usually, though not always, happy in marriage and high in both social and emotional intelligence. But these lucky people are rarely saints, television stars, charming politicians, or glittering celebutantes. Their distinctive quality is not the ability to give a great party or to sway the masses, but an element of warmth, openness, and generosity that draws others in. They are far more likely to be found helping out at their kid's school or going the extra mile at work than blowing past the velvet rope surrounded by paparazzi. Most important, in their inherent abilities, these lucky people are not all that different from any of the rest of us.

The secret to gaining access to social connection and social contentment is being less distracted by one's own psychological business—especially the distortions based on feelings of threat. When any of us feels connected, the absence of social pain and the sense of threat allows us to be truly *there:* in sync with others. This lack of negative arousal leaves us free to be more genuinely available for and engaged by whatever real connection might develop. If a feeling of connection biases cognition, it is in a positive and generous direction that lifts us up while also giving a boost to others. Being socially contented will not necessarily make us the life of the party, but such a generous and optimistic influence often means that others will find us more pleasant and even more interesting.

One of our most intriguing findings about feeling socially satisfied is that this disposition, free of social pain and the distorted social cognitions such pain can cause, also places the individual on a very even—and very healthful—keel. When we feel connected we are generally less agitated and less stressed than when we feel lonely. In general, feeling connected also lowers feelings of hostility and depression. All of which can have profoundly positive influences on our health.

Just as social connection helps keep our entire physical apparatus operating more smoothly, self-regulation—the sum total of an individual's mental and physiological efforts to achieve balance—actually extends outward to other people. A well-regulated, socially contented person sends social signals that are more harmonious and more in sync with the rest of the environment. Not surprisingly, the signals he or she receives back are more harmonious and better synchronized as well. This rippling back and forth between the individual and others is the corollary to self-regulation that we call co-regulation.[17]

In the pages that follow, I am going to delve more deeply into self-regulation, co-regulation, and many other genetic and environmental forces that influence our experience as social beings. To make the benefits—as well as the urgency—of social connection more compelling and more accessible, I am going to examine the tangible consequences of both social pain and social contentment, along with their scientific underpinnings. I want to demonstrate the many ways in which loneliness serves as a new window on who we are as a species. I want to use our recent research findings, woven together within an evolutionary framework, to start trying to change our culture's lopsided view of human nature, its focus on the individual in isolation as the proper measure of all things. But my more immediate goal is to help the socially satisfied get from good to great, while at the same time helping the lonely regain control of their lives. It is my belief that, with a little encouragement, most anyone can emerge from the prison of distorted social cognition and learn to modify self-defeating interactions. What feels like solitary confinement, in other words, need not be a life sentence.

CHAPTER TWO

variation, regulation, and an elastic leash

Between college and graduate school, a young man named Greg moved to New York City, not quite sure what he wanted to do with his life. He came from a background not unlike Katie Bishop's—small-town middle America—and for the first few months he was happy just to be in the Big Apple. In the evenings after work, alone and unencumbered, he would ride the subway to different parts of Manhattan and simply walk the streets, taking in all the sights and sounds. He had a fairly boring job, but still, he felt that he was in the right place and had finally broken free to begin finding out who he wanted to be.

But then Jean, a young woman he had dated senior year, came to town. She needed a place to stay, one thing led to another, and, without really thinking it through, Greg blithely suggested that she could move into his apartment. But the outcome of this impulsive invitation was that a person who had been very happy on his own suddenly found himself with a partner. She was lovely, and he really cared about her, but he soon realized that this was more than he was ready for emotionally. To complicate matters, Jean did not embrace New York with the same relish that he did. She looked to Greg to be her guide and general resource in sorting through all the complications of city living. Then she began to place greater and greater demands not just on his time but on his sense of commitment. After a few weeks, new linens, tableware, and small kitchen appliances began turning up in the apartment. A few more weeks and she began to talk about marriage. Clearly, there was a mismatch between what each of them wanted and expected from their relationship. This forced Greg to come clean about how he had screwed up in suggesting that they move in together, which prompted a huge scene during which she tore into him for his shallowness and immaturity. He had no defense against her accusations of recklessness. He felt guilty as charged, and her harsh words filled him with shame. But then she said how much she loved him. She had no real friends in the city, she still hadn't found a job, and she simply could not face the humiliation of heading back home. She wanted to stay together.

Overwhelmed and confused, Greg could not summon up a constructive plan of action, and he simply went limp. For weeks he sleepwalked through his job, and when he came home he had nothing to say. He was anguished and depressed, but perhaps the worst of it was the profound sense of loneliness that consumed him. He had trouble enough expressing his feelings, and after his big confessional moment, trying to talk further about their situation seemed pointless. He was estranged from his parents, and he was too ashamed of the fix he was in to open up to anyone else.

Then one night he went to meet Jean after her dance class at a theater in Greenwich Village. While he waited for her in the lobby, just inside the plate glass window, he stared out at the street people congregating on the steps of the building across the way. He was feeling the burden of his conflicted emotions as well as the practical problems he had created, and these dilemmas grew in his mind until it seemed as if he were utterly and hopelessly trapped. For a moment he saw his reflection in the glass, and the long and mopey expression on his face looked pretty grim. Then he realized, "That's not my reflection—that's another person looking back at me." One of the street people, raggedy and unshaved, had seen Greg staring out and looking despondent and had squared off in front of him on the other side of the window. What Greg had thought was his own reflection was actually someone else volunteering a spot-on imitation of "self-absorbed young man looking miserable." The fellow on the other side of the glass then jumped back with a wide-eyed, open-mouthed look of surprise, and even Greg had to laugh. That brief moment of human connection not only penetrated his gloom, it penetrated his self-imposed confinement. After that encounter he began to re-engage, and slowly he began to sort out the mess he had created for Jean and himself.

Katie Bishop's minimum daily requirement for social connection was genetically biased for "high." By the time she was grown up she knew she needed something different—she was simply a little off base about what the right "something different" should be.

Greg's genetic thermostat for connection was set much lower. He actually liked being on his own. Nonetheless, he too got stuck for a while in loneliness, not because of a dramatic change in locale, but because of a dramatic mismatch in his intimate social environment.

The famed evolutionary biologist Edward O. Wilson describes the genes as providing a "leash" on our behavior, but a leash that is highly elastic. Our genetic inheritance imposes certain constraints, but it also allows considerable wiggle room. When parents boast about a child's talent for music, or sports, or math, or when they bemoan her talent for mischief, they often wonder about the relative importance of these two major influences—DNA and the world around us. In academic circles, and because of its implications for public policy, this same question has been a topic of heated debate for decades. The psychologist Donald Hebb has compared the query "Which contributes more to personality, nature or nurture?" to the question of which contributes more to the area of a rectangle, the length or the width. The answer is not an either/or, but neither is it a both/and. It is not simply genes *added to* the environment, but genes *interacting with* the environment that typically determines the expression of most basic aspects of personality. The influence of heredity means only that certain individuals, because of their genetic endowment, have a greater need for, or a greater sensitivity to the absence of, connection than others. Whether or not they actually become lonely, either for certain brief periods or throughout their lives, depends on their environment—including their social environment—and environments are influenced by many different factors, including the individual's own thoughts and actions.

To really understand how the genetic influence operates, we need to reach down and examine how we know that genes are involved in the first place. One concept we can be sure of is that nature embraces variety. Conditions in nature are never entirely stable, so each gene pool holds many different options in reserve, which may be why human populations contain both some members whose need for closeness is less intense and others whose focus is much more empathic and socially attuned. The natural virtues of diversity also suggest that, at the level of the individual, one way of being is no better than the other. Those who are highly vulnerable to sensing disconnection can be socially satisfied, and those low in the need for connection can be lonely. The problems arise simply when there is a mismatch between the level of social connection desired and the level the environment provides.

The Genetic Thermostat

The standard way to sort out the heritable (genetic) component from the environmental component of any human characteristic—including the relative intensity of our appetite for social connection—is to do long-term studies of twins. Fraternal twins occur when two separate eggs are fertilized in the same menstrual cycle and develop together in the womb. Because they come from two different eggs, these pairs are no more genetically alike than siblings born from separate pregnancies: On average, they have fifty percent of their genes in common.

On the other hand, identical twins occur when one egg splits in two after being fertilized, giving rise to two embryos early in the developmental process. Barring anomalies, identical twins are for all intents and purposes 100 percent genetically the same.

Between 1991 and 2003, Dorret Boomsma of the Free University Amsterdam asked thousands of identical twins, tracked by the Netherlands Twin Register, to rate how applicable certain statements were to their lives. Two of these statements served as a pretty straightforward gauge of loneliness: "Nobody loves me" and "I feel lonely." Examining how each of the Dutch twins had responded to these statements over the years, we found that the individuals who felt lonely at the beginning of the study tended to feel lonely two or six or even ten years later. Those who felt socially contented and secure at the beginning of the study likewise tended to roll along with feelings that remained roughly similar. But beyond this relative stability, when we found loneliness in one member of a pair of these identical twins, our prediction of loneliness in the second member of the pair was right approximately forty-eight percent of the time.

To understand the significance of that predictive ability, consider that whether an individual is lonely or not lonely is a fifty-fifty proposition, like a coin toss. But our forty-eight percent predictive accuracy was not based on a one-shot flip of the coin. This was a case of being able to predict the outcome forty-eight percent of the time *throughout the whole series of several thousand twin pairs.* The odds of being right by chance half the time in a series of thousands of fifty-fifty propositions are virtually nil. So what our ability to predict indicated was the influence of heredity. It so happens that a heritability coefficient around .48 also holds true for most of the other genetically influenced major personality characteristics such as neuroticism, agreeableness, and anxiety.

As E. O. Wilson's image of the elastic leash suggests, heritability in human behavior means that the genes set the course, but that the environment still strongly influences the final destination. The influence of genes on a purely physical characteristic such as eye color is generally 100 percent, as is the influence of genes on certain conditions such as Huntington's disease. In those cases, genes are in fact destiny; the environment is never going to change the outcome. With the genetic bias toward a greater need for feelings of connection, however, a genetic contribution of forty-eight percent allows for a fifty-two percent contribution from the world we encounter.

Mediating between the genes and the environment is the organism—which is to say, you and me. And here is where social cognition—our subjective perceptions—plays such an important role in determining the net outcome.

Receptivity and Resilience

A few years ago I was asked to give a talk at a scientific conference in Granada, Spain. The invitation came at a particularly busy time in my life, but I was truly enjoying my work, feeling optimistic, and I had always looked forward to this particular meeting. So I shrugged off the inconvenience of one more trip and one more commitment and asked my secretary to book the flight. Little did I know that I was embarking on an unintentional demonstration of how the mind plays gatekeeper to social connection and its benefits.

On the day of the trip I picked up my tickets on the way out the door, rushing as always, and when I got to the airport I discovered that I had a connecting flight through Miami. I thought, this can't be right. So I went to the airline counter and asked, "Is this the flight to Granada?"

"Yep," said the airline employee. "Leaves in an hour."

I had never traveled from Chicago to Europe by way of Miami before, but I had an armful of papers to grade, so I thought no more about it. I sat down in the waiting area, pulled out the stack of papers, and began poring over them. My absorption lifted only at the call for "final boarding," when I had to hurry through the door and down the jetway.

Four hours later I was in Miami, following other passengers off one plane and onto another, inching down the aisle, trying to find my seat for the second leg of my journey. It occurred to me that this was the smallest plane for a transatlantic flight I'd ever seen, so as I passed the flight attendant I said, "We're going to Granada, right?"

She smiled brightly and nodded her head.

"It says we're getting there at one in the morning?" I persisted. "I thought it took overnight."

"Oh, it's not that far," she assured me.

"Okay." Maybe I was missing something. But I was in no mood to question a major airline. Besides, I was eager to get back to grading my students' papers.

Even so, once I settled into my seat and buckled up, I skimmed another paragraph or two and soon fell asleep. The next thing I knew we were on our final approach, with the landing gear jolting into place beneath me.

After we had taxied to the gate and the crew had opened the doors, I retrieved my laptop, my papers, and my carry-on bag and got off the plane. It was dark and I was still groggy, but the landscape looked enough like southern Spain to suit me. Still, there was something not quite right. I felt an odd glimmer of déjà vu.

I drifted through passport control, climbed into a taxi, and said, "Saray Hotel, por favor."

The driver replied, "You say what mon?"

This accent was distinctive, and it was definitely not Spanish.

Nearly twenty years earlier, on vacation, I had visited St. George, the capital of the island of Grenada. I got a queasy feeling as it dawned on me why everything looked vaguely familiar.

I swallowed hard and asked the obvious question. "Are we in Gre-nay-da?"

The driver said, "Yah, sure mon, what you think?"

Chicago to Miami, transferring to a small commuter jet—those connections had nothing to do with Gra-nah-da, Spain. I had just landed in the Caribbean. Not only was I in the wrong country, I was on the wrong side of the planet.

I bolted back into the airport terminal, hoping against hope to find a return flight that could get me to Europe by daybreak. The next plane to Miami was at six a.m., Caribbean time. There was no way I could get back to the States and then on to Spain quickly enough to make the meeting.

I considered my options for a long and painful moment. Then I looked into the sympathetic eyes of the woman behind the counter and sighed.

"Just get me back to Chicago," I said.

I was exhausted, and I felt like a total fool. I had done many ridiculous things in my life, but landing on the wrong side of the world to deliver a scientific paper was a first even for me. The airline personnel were very generous (they seemed to have been through this kind of thing before). They booked me on the six a.m. flight back to Florida and arranged a hotel where I could spend the night.

During the ride through the quiet streets into town, I thought about trying to reach my colleagues to let them know I would not be coming. I had an obligation to these people and I had to set it right. As soon as I checked in at the front desk and got to my room, I picked up the phone and tried to call Spain. But it was now after two A.M. and the international switchboard had closed down for the night.

First time to travel to the wrong country . . . first time to be an irresponsible no-show at a scientific meeting . . . and there was absolutely nothing I could do about it. I could feel my pulse throbbing in my temples.

But then a calm descended. I realized it was not the end of the world. This was not even a matter of life or death. My colleagues had heard me speak before, and they would not perish for being denied one more chance to see my PowerPoint charts and graphs.

I was beginning to feel better. Still, I was too wound up to sleep, so I decided to go downstairs to the hotel bar to relax and get through a few more papers before calling it a night.

This being the Caribbean, the airline had put me up at a resort hotel—something I had neglected to notice when I first arrived, my mind going a zillion miles an hour. I was pleasantly surprised to find that, even well past midnight, the bar was lively, and there was one group in particular, about a dozen men and women, who were having a great time, bantering back and forth in a good-natured way. I sat a few seats away from them and began grading the next paper, but it didn't take long before one of them asked what I was doing. God knows I must have looked out of place—Super Nerd in the islands. The conversation soon led to the story about how I came to be in a St. George hotel bar rather than on a flight to Spain, and we all had a pretty good laugh about it. It turned out that my new acquaintances were professional footballers visiting from England with their families. I had played football in school and at college—

although we called it soccer—and despite the fact that I had never been very good, I loved the game. So we talked about sports, and then about other things. We were still talking and laughing when I noticed the sky brightening toward dawn and it was time to get back to the airport for my flight home.

The irony, of course, is that the work I do, and the work I was going to report on in Granada, is all about the balm that connection can provide for the stresses of life, even the stress of being stranded far, far from where we are supposed to be. And yet the real point of the story is that the soothing power of social connection depends on having a clear channel to receive it.

Our level of vulnerability to feeling disconnected is in part at the mercy of our genes. The self-regulation that keeps our social receptors free of static can be difficult when the environment does all it can to frustrate our pursuit of what our genes demand. But our thoughts are something we can address directly, which is why we can use social cognition as a leverage point for regaining control of our social experience. The way we think about social situations can prepare us to metabolize the almost medicinal qualities of social warmth, or it can set us up to confirm the cynical aphorism that "hell is other people."

Serving as a prompt to restore social bonds, loneliness increases the sensitivity of our receptors for social signals. At the same time, because of the deeply rooted fear it represents, loneliness disrupts the way those signals are processed, diminishing the accuracy of the message that actually gets through. When we are persistently lonely, this dual influence—higher sensitivity, less accuracy—can leave us misconstruing social signals that others do not even detect, or if they detect, interpret quite differently.

Reading and interpreting social cues is for any of us, at any time, a demanding and cognitively complex activity, which is why our minds embrace any shortcut that simplifies the job. We typically start by forming expectations emotionally, then use our reasoning powers to confirm what our emotions have led us to expect. We do this in forming our first impressions of people and situations, our political opinions—a great many of our preferences.[1] We invariably take cognitive shortcuts, but when we are lonely, the social expectations and snap judgments we create are generally pessimistic. We then use them to construct a bulwark against the negative evaluations and ultimate rejection that the fearful nature of loneliness encourages us to anticipate.

When we feel socially connected, as most of us feel most of the time, we tend to attribute success to our own actions and failure to bad luck. When we feel socially isolated and depressed, we tend to reverse this useful illusion and turn even small errors into catastrophes—at least in our own minds. Meanwhile, we use the same everyday cognitive shortcuts to try to barricade ourselves against criticism and responsibility for our screw-ups. The net result is that, over time, if we get stuck in loneliness, this complex pattern of behavior can contribute to our isolation from other people.

For each of us, throughout our lives, the balance between need and satisfaction can shift, and the pressures on our ability to regulate our emotions can vary. There have been plenty of times in my life when I felt very alone, but luckily, when I took my inadvertent trip to Grenada, I was feeling particularly upbeat and connected. That may be why, even though I knew that missing the meeting in Spain was going to make certain professional colleagues very upset with me, I never imagined that the gaffe would turn me into an outcast. Had this screw-up happened at some more difficult time in my life, a time when I was feeling isolated, personally or professionally, my response might have been very different. I might not have been able to accept and adapt to the silly and quite embarrassing circumstance in which I found myself. At that very nice resort hotel in Grenada I might have tossed and turned all night, reliving the anger and humiliation in my mind. As soon as the phone lines opened in the morning I might have called my secretary and taken her head off, blaming the entire fiasco on her (while denying the contribution of my own considerable lack of due diligence—not checking the spelling of the destination on my ticket, ignoring the clear hints that I was off track). Or, when I went downstairs to the bar, my goal might have been to stew my problems in alcohol. Rather than letting myself be drawn into convivial talk with a group of strangers, I might have avoided conversation for fear that it would expose me to ridicule and social exclusion: "You did what!!??" In truth, everyone at the hotel bar that night did share a laugh about the absentminded professor so oblivious to his surroundings that he let himself be shipped off to Gre-nay-da rather than Gra-nah-da. But being in a good place in terms of my feelings of social connection, I was able to find it pretty funny myself, and it was that relaxed perception that allowed me to get through the experience none the worse for wear.

The Price of Loneliness

When I got back from my unplanned trip to the Caribbean, my secretary and I managed to laugh about the incident as well. We also made a pact that, knowing my more than occasional preoccupations, we would try to at least keep me in the right geographical region. But aside from holding the story of the oblivious professor in reserve for toasts and roasts, we pretty much left it at that.

Loneliness, by contrast, can make us less able to get beyond even the normal disruptions, setbacks, and mistakes of day-to-day life. The inability to let go of such events has, in turn, consequences that are not just social but physiological: Loneliness creates a subtle but persistent difference in cardiovascular function that sets the stage for trouble later in life. This finding, combined with the fact that loneliness can persist and remain stable through the years, means that its negative effects on health, even the subtle ones, have plenty of time to accrue and compound.[2]

For young people, loneliness is not associated with overtly unhealthful behaviors. Among young adults, in fact, alcohol consumption—at least as represented by social drinking—is actually less of a problem among those who are lonely than among those who feel socially contented. By middle age, however, lonely adults consume more alcohol and engage in less vigorous exercise than those who are not lonely.[3] Their diet is higher in fat. They sleep just as much as the nonlonely, but their sleep is less efficient, meaning less restorative, and they report feeling more daytime fatigue.[4]

Even though objective measures suggest that their circumstances are no more stress-filled than those who are socially contented, lonely young people perceive themselves to be having a tougher time, and, over time, the stress of that subjective sense of being under the gun can create wear and tear throughout the organism. By the time they reach middle age, people who are chronically lonely actually do endure more objective stressors than those who are socially satisfied. Middle-aged adults who are lonely have more divorces, more run-ins with neighbors, more estrangement from family. By middle age, the tougher time they may have perceived themselves to be experiencing has become a reality.[5]

But once again, people who get stuck in loneliness have not done anything wrong. None of us is immune to feelings of isolation, any more than we are immune to feelings of hunger or physical pain. The interaction between a genetic bias and life circumstances that constitutes loneliness is generally beyond our control. However, once it is triggered, the defensive form of thinking that loneliness generates—a lonely social cognition—can make every social molehill look like a mountain. When we are lonely we not only react more intensely to the negatives; we also experience less of a soothing uplift from the positives. Even when we succeed in eliciting nurturing support from a friend or a loved one, if we are feeling lonely we tend to perceive the exchange as less fulfilling than we had hoped it would be.

For creatures shaped by evolution to feel safe in company and endangered when unwillingly alone, feelings of isolation and perceptions of threat reinforce each other to promote a higher and more persistent level of wariness. To prepare us to react efficiently when confronting threats to life and limb, nature provided us with the ability to be cognitively hypervigilant, along with a chain of physiological reactions known as the fight-or-flight response. But the neurological wiring we depend on today evolved in response to the kinds of hit-and-run stressors we faced millions of years ago. As a result, our stress response ("fight or flight") includes a prompt to immediate action that increases resistance in the cardiovascular system and floods the body with hormones that rev us up. If we were fending off wild dogs, those hormones could help save our lives. However, when our stressors consist of feeling isolated and unloved, the constant presence of these excitatory chemicals acts as a corrosive force that accelerates the aging process.

Luckily, the grinding effect of stress brought on by a persistent sense of being alone is only part of the story. Our research takes into account the whole constellation of social, psychological, and biological events, including the vitally important counterweight to the fight-or-flight system—what my colleagues and I call the physiology of "rest and digest." Just as our cells and organ systems undergo wear and tear, often as a result of stress, they also benefit from inherent processes of repair and maintenance that are associated with restorative behaviors such as sound sleep. As you might expect from what we have seen so far, some of these maintenance and repair functions of the human mind and body are also heavily influenced by the social world.

A Leash, But Elastic

Molecular biologists once estimated that human DNA contained something on the order of a hundred thousand genes. This seemed to make sense, given the number of cellular processes operating within our physiology, our intricate neural hardware, and the subtle shadings of our infinitely complex behavior. But when they succeeded in mapping out the entire human genome, itemizing all the genetic instructions that, if followed by a developing embryo, will result in a fully functioning human being, they discovered that we have barely more genes than *Caenorhabditis elegans* (a worm), and only half as many as *Oryza sativa* (the cultivated rice plant). The revised number of genes in the human genome turned out to be around thirty thousand.[6] More recently, that number was revised downward again, to a range between twenty and twenty-five thousand.[7] The distinctive complexities that make us human depend not so much on the number of genes, but once again, on the ways in which those genes interact with one another, and with the world around us, through the sensory, integrative, and motor systems the genes control. Which seems entirely appropriate for a creature so dependent on interactive social connections and so skilled at adapting to novel environments.

Relative to body size, the amount of gray matter in the human brain as a whole—even the amount in the prefrontal cortex, the part involved in self-regulation and other executive functions—is barely larger in humans than in our cousins the apes.[8] We have more cortical neurons for our size than other mammals, but barely more than whales and elephants.[9] Our cognitive advantage lies in a combination and enhancement of properties that already exist in our nearest relatives: the chimpanzees, and a closely related species, the bonobos. Having more intelligence has adaptive value for large mammals because it facilitates discovering better ways to find or capture food, avoid perils, and navigate territories, but the complexities of these demands pale by comparison to the complexities of social living. Living in groups placed a premium on having the ability to recognize the mental states of others—a capacity called "theory of mind." But once again, theory of mind is a form of social cognition, an ability that becomes readily distorted through the experience of loneliness.

But There's a Catch

Whether you are a relatively independent Greg or a need-to-be-close Katie, no one wants to feel the pain of loneliness, and no one should be blamed for being trapped inside it. What makes loneliness especially insidious is that it contains this Catch-22: Real relief from loneliness requires the cooperation of at least one other person, and yet the more chronic our loneliness becomes, the less equipped we may be to entice such cooperation. Other negative states, such as hunger and pain, that motivate us to make changes to modify unpleasant or aversive conditions can be dealt with by simple, individual action. When you feel hungry, you eat. When you feel a sharp pain in your lower extremity, you move your toe off the tack. But when the unpleasant state is loneliness, the best way to get relief is to form a connection with someone else. Each of the individuals involved must be willing to connect, must be free to do so, and must agree to more or less the same timetable. Frustration with the difficulty imposed by these terms can trigger hostility, depression, despair, impaired skills in social perception, as well as a sense of diminished personal control. This is when failures of self-regulation, combined with the desire to mask pain with whatever pleasure is readily available, can lead to unwise sexual encounters, too much to drink, or a sticky spoon in the bottom of an empty quart of ice cream. Once this negative feedback loop starts rumbling through our lives, others may start to view us less favorably because of our self-protective, sometimes distant, sometimes caustic behavior. This, in turn, merely reinforces our pessimistic social expectations. Now others really are beginning to treat us badly, which seems like adding insult to injury, which spins the cycle of defensive behavior and negative social results even further downhill.

But the real injury added to the insult is that, while this outward disruption is taking place, loneliness also disrupts the regulation of key cellular processes deep within the body. This is how chronic loneliness not only contributes to further social isolation but predisposes us to premature aging. Chronic loneliness not only makes us miserable, then, it can also make us sick.

CHAPTER THREE

losing control

Getting by as a happy and healthy human being requires the integrative intelligence exercised by the brain's frontal lobes, a function that neuroscientists and psychologists have labeled executive control. Remembering your name does not require this kind of intellectual coordination and integration, nor does simple arithmetic. Certain other tasks, such as reading your native language or playing a piece on the piano, you readily push out of executive control once you've mastered them. But more complex cognitive functions, including the complexities of social behavior, demand lifelong self-regulation. It is these social cognitions and behaviors that go haywire when our sense of belonging takes a hit.

While I was teaching at Ohio State in the 1990s, we wanted to measure the effect of loneliness on the ability to focus and maintain attention. We used the UCLA Loneliness Scale to sort undergraduate volunteers into three groups: those who felt very lonely, those who felt not at all lonely, and those who were somewhere in the middle. Then we subjected all three groups to a cognitive test called dichotic listening.

Our experiment relied on the fact that the human brain has two halves, or hemispheres, which maintain a division of labor called lateralization. This means that one side or the other takes the lead in regulating specific functions. Much of our comprehension and production of language, for instance, is governed by the left hemisphere. To a large extent, spatial reasoning, as well as variation in tone, cadence, and pitch when speaking, is governed by the right. An added twist is that, when it comes to sensory perception and motor skills, the left hemisphere controls the right side of the body and the right hemisphere controls the left. The *left* side of the brain is slightly dominant in most people, which means that most people are slightly stronger and more dexterous on the *right* side of the body. Similarly, most people—being *left*-brain dominant for language—are better able to focus on and detect the fine points of verbal information presented to the *right* ear. Usually, this general right-ear advantage can be overcome when we ask the volunteers in experiments to concentrate on the sounds coming to the left ear. In other words, by exerting a great deal of executive control—the integrative intelligence exercised by the brain's frontal lobes—the volunteers can usually override the natural bias that favors the right ear.

For consistency in our study of the effect of feeling isolated on dichotic listening, we selected only right-handed (left-brain dominant) students from all three groups (high, low, and "normal" levels of loneliness). We asked them to put on headsets that allowed us to separate and control the sounds streaming into each ear. Their task was to identify the sound (a consonant-vowel pair such as "wu" or "ha") being presented to one ear while irrelevant word sounds were presented to the other ear. In general, when not given any instructions about which ear to attend to, the students showed the right-ear advantage—their accuracy was better in identifying the sounds coming into that side. When instructed to attend to the right, or dominant ear, all three groups did so equally well. When we asked them to focus their attention on the left ear—to consciously override the normal processing of sound that is biased toward the right ear—those from the "normal" group and those from the socially connected group still did pretty well. The students identified as lonely, however, were less successful at imposing conscious control. They were less accurate than the other groups in tuning out what they heard in the right ear in order to focus on and identify sounds heard in the left. Loneliness, in effect, had given these individuals an attentional deficit—at least with regard to this one, rather difficult task.[1]

Tuning In, Tuning Out

Just as diabetes disrupts our ability to self-regulate the level of sugar in the "internal environment" of our blood, loneliness can disrupt our ability to self-regulate in the external, social environment. As we will see in numerous examples to come, the inability to use self-regulation to tune out distraction and focus the mind often gets in the way of our attempts to connect with others when we feel lonely. Other problems in self-regulation specifically attributable to loneliness have turned up in measures of alcohol abuse, drug abuse, bulimia nervosa, even suicide.[2] But long before feelings of isolation manifest themselves in these serious health problems, impaired self-regulation causes lonely individuals every day, everywhere, to act in ways that, sadly, do nothing more than reinforce their loneliness.

While the objective in going to certain bars and dance clubs appears to be getting drunk and hooking up, how many of the people crowding in are actually driven by a deeper craving for human connection that they simply don't know how to pursue? That they might fail to find truly satisfying connection amid blaring noise and shouted conversation—often interrupted by someone's cell phone—is not entirely surprising. Unfortunately, their failure to find what they need then makes them all the more susceptible to the slightly out of control behavior that often begins in bars and dance clubs.

Being able to self-regulate our emotions and our behaviors is a large part of what makes us human. The extent to which we rely on executive control to be an effective member of human society intensifies the tragedy when stroke, infection, or injury impairs the brain systems responsible for that control. The classic example is Phineas Gage, a ninteenth-century railroad worker who had a one-meter-long (but mercifully slender) steel rod blown through his brain in a construction accident (see Figure 3). An hour after the accident he was chatting about the experience. Within a couple of months he had recovered from his obvious injuries. But, according to various accounts, he was no longer the responsible, personable young man, or the serious and conscientious young supervisor, he had been. He had become a foulmouthed hothead, incapable of holding a job or planning his future. His doctor described him as "fitful, irreverent, indulging at times in the grossest profanity, which was not previously his custom, manifesting but little deference for his fellows, impatient of restraint or advice when it conflicts with his desires, at times pertinaciously obstinate, capricious and vacillating . . . a child in his intellectual capacity and manifestations, he has the animal passions of a strong man."[3]

Despite the fact that his physical and general reasoning capabilities—attention, perception, memory, language, and intelligence—were intact, he could no longer make good choices. Because he could no longer incorporate social conventions and ethical concepts into his social interactions, his decisions no longer took into consideration the concerns of others, and, consequently, his decisions no longer served his own, or anyone else's, long-term interest.[4]

A Stressed Executive

Phineas Gage died penniless twelve years after his accident. I doubt that loneliness alone has ever accounted for such a dramatic disruption of personality, but the psychologists Roy Baumeister and Jean Twenge have demonstrated that feeling socially excluded can get in the way of our exercising some of the human characteristics we value the most. These researchers provided the first experimental proof of what anyone who ever went to middle school might reasonably suspect—that feeling left out can reduce executive function sufficiently to impair mental performance.

To study the effects of social disconnection on executive control, Baumeister and his team rounded up undergraduate volunteers and asked them to complete two questionnaires: an introversion/extraversion test and what they described as a personality inventory. The team gave the volunteers a reassuringly accurate report on the results of the introversion/extraversion test, but only to bolster the students' level of comfort and confidence. This ploy was necessary because the feedback on the personality inventories they were about to provide was entirely bogus.

At times, social science research may seem like one of those reality shows with a hidden camera that traps unwary people in ridiculous, almost sadistic, situations. "You're lucky," the Baumeister group told some of their volunteers. "You're the type who'll have

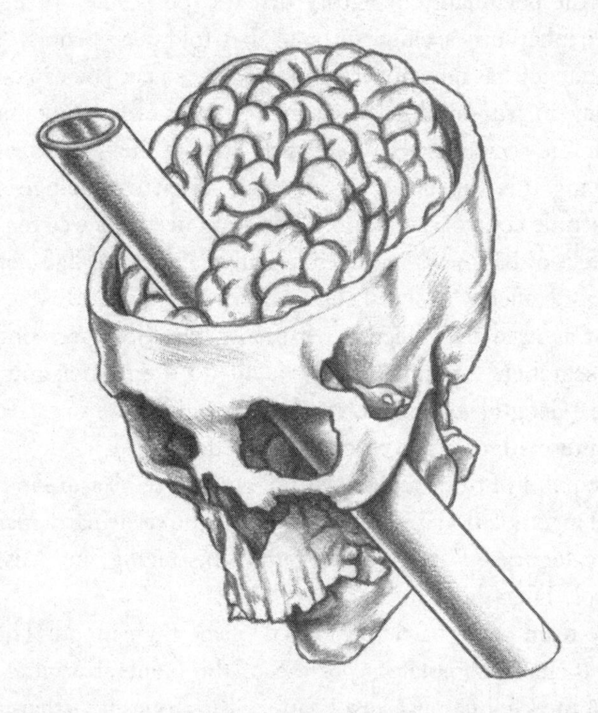

FIGURE 3. The path of the steel rod through Phineas Gage's head. It went through his skull and eye and destroyed portions of his orbitofrontal and ventromedial prefrontal cortex. Adapted from H. Damasio, T. Grabowski, R. Frank, A. M. Galaburda, and A. R. Damasio, "The return of Phineas Gage: Clues about the brain from the skull of a famous patient," *Science* 264, no. 5162 (May 20, 1994): 1102–5.

rewarding relationships throughout your life. Most likely you'll have enduring friendships and a long and happy marriage, with plenty of people who'll always care deeply about you."

On others, they dropped a fairly large psychological bomb. "Hate to say this, but according to these results, you're the type who probably will end up alone. You may have friends and relationships now, but by your mid-twenties most of these will have drifted away. You might get married . . . Actually, you may have several marriages, but they're all likely to fail. Certainly they won't continue into your thirties. Relationships just won't last for you. Odds are, you'll end up more and more alone the longer you live."

For members of a third group, the research team provided a reading of the personality inventory that was purposely off the point. "You're inherently accident-prone," they told these people. "Even if this tendency has not shown up in your life so far, you can count on breaking an arm or a leg fairly often, perhaps even being injured in automobile accidents very frequently later in life." The reason for conveying this bad news was to create what psychologists call a "misfortune control condition." The team needed a way to sort out the effect of bad news in general from the effect of bad news pertaining specifically to social connections.

Rest assured that when experiments like these are completed, the researchers explain all these details to the participants, going to great lengths to ensure that none will come away from the encounter with negative psychological effects.

The point of this exercise was not a perverse pleasure in psychological manipulation, but the need to sort the study participants into three categories: Future Alone, Future Belonging, and Misfortune Control. Then the real test began.

The team asked participants to describe their mood. Then they asked them to complete a portion of the General Mental Ability Test of the Graduate Record Exam, a list of questions that includes measures of mathematical ability, spatial ability, and verbal reasoning.

In describing their mood, those who had been given the bad news about their long-term social life showed no indication of emotional distress, suggesting that their decline in cognitive ability was not a simple matter of being flustered. Even so, the Future Alone group performed significantly worse on the exams than the Future Belonging group, those who had been told that their future would be socially rosy. The Future Alone group showed impairment in both speed and accuracy.[5] This group also scored significantly worse than the Misfortune Control group, those who had received dire predictions of physical calamities, rather than social ones. Bad news itself, then, was not enough to cause the disruption in mental abilities, only bad news about social connection.

Even transient feelings that one is likely to face the future alone, then, can increase the difficulty of self-regulation, undermining, as in this case, the ability to think clearly. (Don't tell your eighth grader, but while not getting invited to a party is not an *excuse* for doing poorly on an algebra test, it may be a valid *explanation*.)

But Baumeister and his colleagues wanted to dig deeper to explore more fully the range of cognitive performance that might be vulnerable. To achieve this more subtle analysis, they did further studies with slight variations.

In one of these, the three groups of participants completed the mood measure and then some members of each group were assigned to complete a logical reasoning task, while others were assigned a rote memorization task. When it came to rote memorization and recall—the simple stuff—the Future Alones did no worse than anyone else. But on logical reasoning, the Future Alones attempted to solve the fewest problems and made the most mistakes with those they did attempt. They performed significantly worse than the Future Belongings and the Misfortune Controls. The prospect of a future of social isolation did not impair routine mental ability, only the higher-order processes that integrate and coordinate. Indeed, in another study, a brain scan conducted while subjects performed moderately difficult math problems confirmed that the brains of the socially excluded participants were less active in the same areas related to executive control that, as in the dichotic listening task, allow us to maintain focused attention.[6]

The Lifelong Challenge

Maintaining control is a challenge we confront all our lives, and not only when facing math problems or perplexing experiments with sound. As infants, we all dwell in the land of immediate gratification and self-interest, but over time, and under appropriate parental influences, we learn not to grab toys away from our playmates, hit our siblings, or eat an entire box of cookies twenty minutes before dinner. Our facility with regulating our impulses improves vastly as we mature, but regulation still requires control from the top, especially when we have to decide whether to exercise self-restraint: whether to spend or save, eat what we please or what's good for us, see what's on TV or do our taxes.

Baumeister and his team wanted to find out if the prospect of a future alone would affect these "self-restraint" aspects of self-regulation in the same way it affected cognitive abilities. One measure would be the ability to persist with a task that provided a beneficial long-term outcome but was seriously unpleasant. Once again, through bogus feedback that manipulated feelings of disconnection, the researchers sorted participants into Future Alone, Future Belonging, and Misfortune Control groups. Then they assigned each participant the task of drinking a concoction that, as they clearly explained, was healthful and nutritious but had an unpleasant taste. They measured the number of ounces each participant was willing to drink in return for a small monetary reward. Results followed the same pattern as before—the Future Alone participants were far less persistent at the task than those who had not been socially undermined. The Future Belongings were far more willing to endure a little bit of unpleasantness in return for a payoff.[7]

Baumeister and his team devised a follow-up study to test the flip side: How well would participants be able to self-regulate in limiting behavior that felt good but was clearly not good for them? They assembled a new group of volunteers and told them that the study would involve forming small teams of people who liked and respected each other. A researcher explained that, after having a few moments to make friends and mingle with the whole group, each participant would be allowed to pick the two people with whom he or she would most like to work. In one-on-one follow-up sessions, the researcher then feigned embarrassment as she told certain participants that no one had chosen them. "But that's fine," she said. "You can just go ahead and complete the next part of the task alone." The other participants were given a signal from the other end of the pleasure/pain continuum. They were told that *everyone* had expressed the desire to work with them, but that because there could not be a group so large, they too would have to complete the next part of the task alone. So those in one group had just been given the social equivalent of a poke in the eye, while the others had been given a big warm hug.

The task to be completed was framed as a taste test. Each participant, both the socially disconnected and the socially embraced, was given a bowl of thirty-five bite-sized chocolate chip cookies. They were instructed to "test" the cookies by eating as many as they needed to make an accurate judgment of taste, texture, and smell. (As a control measure, interviews with other students from the same university confirmed that eating cookies was almost universally considered bad behavior with unhealthful consequences.)

Participants who had been set up to feel socially disconnected consumed, on average, nine cookies—roughly twice as many as those eaten by the participants who had been told everyone wanted to work with them. Social disconnection not only whetted the appetite for fattening food, it appeared to make the cookies taste better: Most of the participants who had been primed to feel excluded rated the cookies more favorably on taste than did the socially accepted tasters. But seemingly as a testament to a failure in self-regulation, they ate a relatively high number of cookies even when they did not find them particularly appealing. Many of the socially disconnected volunteers who ate the most cookies still rated them as being mediocre at best.

Is it any wonder that we turn to ice cream or other fatty foods when we're sitting at home feeling all alone in the world? We want to soothe the pain we feel by mainlining sugar and fat content to the pleasure centers of the brain, and, absent self-control, we go right at it. This loss of executive function also helps explain the oft observed tendency of rejected lovers to do things they later regret.

A man interviewed by the sociologist Robert Weiss found himself compulsively driving past the house he had stormed out of only weeks before. Each time he did this, he said, his edginess somehow subsided. A woman found herself having to fight the urge to call her ex to talk about her pain, despite his considerable contribution to that pain, and also despite the fact that she had no desire to get back together with him.[8] In both cases, we see dysregulated, lonely individuals struggling to make themselves feel better, if only for the moment.

Certain behaviors can succeed in regulating mood, just as taking a drug or a drink can regulate mood. A primary benefit of having satisfying social connections is that it allows us to find self-regulatory behaviors that promote resonance with others (rather than humiliation and regret), that do not put us in awkward or dangerous situations, and that are not, ultimately, counterproductive when it comes to enhancing the satisfaction we find in social connection.

As a sociologist, Weiss observed that feeling excluded increases a person's motivation to make new friends, to create a positive impression on unfamiliar others, to work with others, and to view others more favorably than they might actually merit. Each of these behaviors has been confirmed by more recent investigations.[9] But when the desire to affiliate becomes thwarted, prolonged feelings of social disconnection turn the positive impulses toward the negative. In one experiment, participants made to feel excluded evaluated others more harshly, and when the rules of the experimental exercise called for sanctions, they agreed to administer more punishment (a painful blast of noise) to fellow participants.[10] Those who felt excluded were also less willing to donate money to a student fund, or to offer to help a stranger following a mishap staged as part of the experiment. They were also more inclined to take irrational, self-defeating risks, and to procrastinate, indulging themselves with pleasurable tasks when they needed to be studying for upcoming tests.[11]

Baumeister's study subjects had been jolted into feelings of disconnection that would quickly pass. When the experiment was over, they were reassured that the forecast of a future of social isolation had not been legitimate, given a small payment to compensate them for their time, and sent on their way. But what about those for whom loneliness persists as a real-life issue? How does feeling isolated, day to day, every day, affect executive control?

A Condition That Does Not Improve with Age

In 2002 our team at the University of Chicago began collecting longitudinal data on a representative sample of middle-aged and older citizens in the greater Chicago metropolitan area. We subjected these volunteers to numerous physiological and psychological measurements, including the UCLA Loneliness Scale. The longitudinal approach allowed us to track changes over time, and focusing on older citizens also extended our data to a study population beyond the limited universe of healthy young people—college undergraduates—that generally provides most of the volunteers for this kind of research. We also selected middle-aged and older adults for study because this is a period in which baseline bodily functions deteriorate, making it a kind of critical period for investigating the possible effects of loneliness on health.

When we analyzed the diets of these older adults, what they ate week after week, month after month in real life—then statistically controlled for irrelevant factors—our finding dovetailed with Baumeister's results from his one-time-only cookie tasting. Just like young people made to feel disconnected in the moment, older adults who felt lonely in their daily lives had a substantially higher intake of fatty foods. In fact, we found that the calories of fat they consumed increased by 2.56 percent for each standard deviation increase in loneliness as measured by the UCLA Loneliness Scale.[12]

In another study, after giving his college-age participants expectations of a future characterized by social isolation, Baumeister asked them to trace a geometric figure without double-tracing any lines and without lifting pen from paper. This appeared to be a garden-variety test of spatial reasoning, but in fact the puzzles had been rigged so as to be unsolvable. The real test was to see how long each participant would stick with a task that was—though they didn't know it at the time—going nowhere.[13] Those who had just been told that they could anticipate a future without meaningful relationships gave up significantly sooner than participants in all other conditions. When we gave our lonely older folks from Cook County psychological inventories designed to measure self-regulation, they too showed that they were less able than their more socially contented peers to persist in the face of adversity.

Having to cope with loneliness when your persistence is impaired by loneliness seems awfully unfair, and interviews with lonely people from real life illuminate the pain of the ongoing struggle. Some try to impose self-regulation by instituting little routines: "I find that the lonesome part of the day is at dusk," one woman said. "I don't have my mail delivered and I usually go out and get my mail or do some shopping at this time—just to get away from the house."[14]

Such efforts can be simple, ritualistic tasks—the kind that do not require much in the way of executive function. But sometimes pleasant routines can devolve into busywork. The more fortunate find activities to regulate their behavior that can become more than that, new routines that actually provide meaning: "And if I haven't got anything else to do I'll make a pie for somebody, one of the neighbors that I hear is going to have company . . . Make cookies, give them away . . . there's always places we can take pounds of cookies—orphanages . . . You come home and you think, '. . . I did something,' you know. The space you're occupying counts for something."[15] In reaching out to do something for others, this woman was definitely taking a step in the right direction. For those whose loneliness has become a more persistent and disruptive condition, however, finding a solution may require making more fundamental changes in how they see their social world.

Regain Your Grip, Reframe Your Cognition

In the early 1990s the psychologists Sally Boysen and Gary Berntson led the way in studying the self-regulatory abilities of mathematically adept apes. Boysen had trained a chimp named Sheba to understand and use Arabic numerals. In 1996 I joined Boysen and Berntson in a study of primate preferences that put Sheba's skills on display.[16] In one experiment, we asked our study participant to choose between two plates, each baited with a different amount of candy ranging from six pieces down to none. When she first saw the two plates behind a glass partition (which was there to keep her from simply grabbing whatever she wanted), she immediately pointed to the one with the larger number of candies. No surprise—like most of us, Sheba had a sweet tooth. However, the name of the game we wanted her to play was "reverse contingency." The rule she had to learn was that she would get the number of candies on the plate to which she did *not* point. To get the six candies, in other words, she had to point to the plate with only three.

Sheba had a very tough time resisting the urge to point to the larger array, and the same was true of other chimps we tested. Overall, when their goal was tantalizingly right there in front of them, they followed the rules only thirty percent of the time, so seventy percent of the time they were skunked with the smaller payoff.

But then we made a switch and played the same game with the numerals 0 through 6 instead of the candy itself. The reverse contingency remained the same: If Sheba pointed to the numeral 3 instead of the 6, we would give her the larger number of candies. When the task involved numerals only—symbolic representations, no candy in view—Sheba did dramatically better. As a group, she and her fellow simian mathematicians doubled their rate of success, getting it right sixty-seven percent of the time.

In confronting a large pile of candy to which they should *not* point, our chimps encountered what psychologists call an "interference effect." In the dichotic listening study described earlier, the interference took the form of sound coming into the dominant ear when the task was to attend to what was coming into the other ear. With the chimps, the source of the interference was the overwhelming appeal of candy in plain sight. Both when working with the numbers and with the actual candies, the chimps knew what we were asking them to do. They had learned the rule, and they knew their number concepts. They simply could not help being distracted by the sight of all that sugar.

Our lonely study participants also knew the rules of the dichotic listening task—they simply could not muster up the degree of self-regulation necessary to overcome the distraction. Similarly, people who feel lonely know better than to gorge on ice cream, berate co-workers, sleep around, or yell at their new husband for bringing home the wrong kind of jelly. We humans simply have more difficulty overcoming these impulses when we feel isolated than when we feel connected.

In another study, researchers asked participants either to describe a personal problem to an assigned partner, or to adopt the role of listener while the partner described his or her problem.[17] Lonely individuals, when specifically requested to take the helping role, were just as socially skilled as the others. They were active listeners, they offered assistance to their partners, and they stayed with the conversation longer than those who were describing their troubles. So we retain the ability to be socially adept when we feel lonely. But in the conditions of real life, as opposed to experimental conditions when we are asked specifically to play a certain role, loneliness prevents us from putting that ability to good use. This failure to put our best foot forward is compounded by lack of confidence and general negativity: Despite their display of skill in the experiment, the lonely participants consistently rated themselves as being less socially adept than other people.

The most useful lesson we learn from the chimps is this: If you want to reestablish the self-regulation you need in order to benefit from the knowledge and the skills you have, reframe the stimulus. With the chimps, when we moved from the appetitive to the abstract by introducing symbolic representations—that is, numbers instead of actual candies—they doubled their ability to maintain control.

Being humans rather than chimps, each of us has the remarkable capacity to reframe representations of stimuli within our own heads. This is both imminently doable, and, of course, easier said than done. It takes practice, and the primary difficulty of successful reframing is rooted in the peculiar architecture of our brain. Humans have more complex brains than apes, with new capacities that emerged very late in our own distinct evolutionary progression. These new features allow us to, among other things, manipulate numbers far beyond the range from zero to six. These new capacities, located in the part of the brain called the neocortex, allow us to write symphonies, discuss which actor made the best James Bond, and plan trips to Mars. However, all the more recent and more sophisticated wiring that makes us so smart did not replace the much older and more primitive neural processes that we share with monkeys or even mice. The older systems are still running beneath the newer ones, and often quite independently. In computer jargon, this progressive layering of function would be called a layered upgrade rather than a download and overwrite. In psychology it is called a re-representation, and it is distributed widely across different levels of the spinal cord and the brain. What this layered upgrade means, in practical terms, is that the neocortex is not the undisputed captain of the ship. The neocortex is up on the bridge, observing and aware, planning and making decisions, but there is always grumbling below deck from the more primitive and emotional layers of the brain that were on board long before the neocortex showed up (see Figure 4).

Later on we will explore how the brain copes with this complexity within a social context, but what we've seen so far already suggests why a lonely person cannot simply take off her glasses as they do in the movies, get a new hairstyle, and become the belle of the ball. It was not conscious decisionmaking or even negative mood that caused study participants who had just been socially excluded to

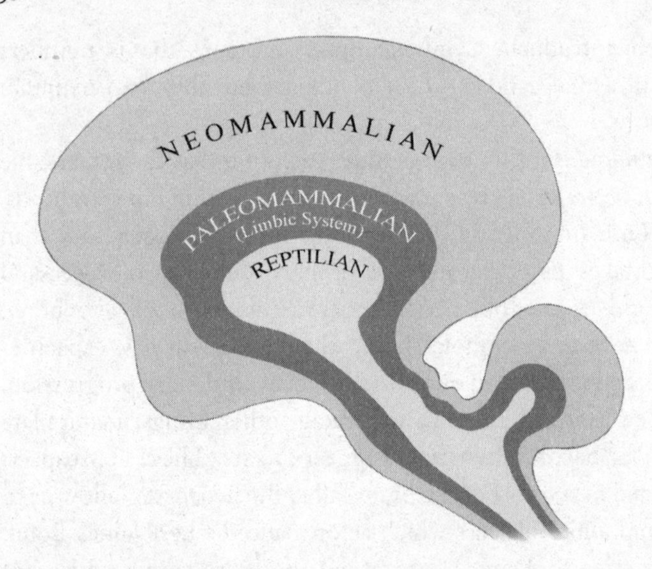

FIGURE 4. The triune brain, a model proposed by the neurologist Paul MacLean in the early 1950s to describe the evolution of the structure of the brain. According to MacLean, the oldest part of the brain, the reptilian brain, consists of the lowest portions of the brain as it emerges from the spinal cord (e.g., brain stem, medulla, pons, cerebellum) and controls instinctual survival behaviors, autonomic functions (e.g., blood pressure, breathing), and balance. The paleomammalian or limbic brain (e.g., hypothalamus, amygdala, hippocampus) evolved next; it controls responses to basic motivations and emotions such as feeding, drinking, fighting, fleeing, and sexual reproduction. The neomammalian brain, also known as the cerebral cortex (e.g., cerebrum, cortex, neocortex), evolved most recently and controls higher-order processes such as thinking, reasoning, language, problem solving, emotional regulation, and self-control.

eat more of an unhealthful food, drink less of a beverage that was lousy tasting but good for them, give up more quickly on a frustrating task, and be less successful at screening out distractions to focus on the business at hand. And it was not having heard disturbing predictions in general—the prospect of a future filled with broken bones and painful injuries—that threw people off their game. In the studies I've just described, there was only one force that could impair self-regulation enough to disrupt both thinking and behavior. This disturbing, dysregulating force is a fear rooted both in each individual's earliest moments of life, and in the earliest moments of our history as a species. That overwhelming fear is the terror of feeling helplessly and dangerously alone.

CHAPTER FOUR

selfish genes, social animals

If you asked a zookeeper to create a proper enclosure for the species *Homo sapiens*, she would list at the top of her concerns "obligatorily gregarious," meaning that you do not house a member of the human family in isolation, any more than you house a member of *Aptenodytes forsteri* (Emperor penguins) in hot desert sand. It simply makes no sense to put a creature in an environment that stretches its genetic leash quite that far.

Nonetheless, for five centuries or more—and at a much quicker pace during the past five decades—Western societies have demoted human gregariousness from a necessity to an incidental. In fact, the latest figures show that ever-greater numbers of people are accepting a life in which they are physically, and perhaps emotionally, isolated from one another. Consider this sampling:

- Respondents in a 2004 social science survey were three times more likely to report having no one with whom to discuss important matters than respondents in 1985.[1]

- During the past two decades, more or less, the average household size in the United States declined by about ten percent, to 2.5 persons.

- In 1990 more than one in five households with children under eighteen was headed by a single parent. Currently, the proportion of single-parent households is nearly one in three.

- In the United States in 2000 there were more than twenty-seven million people living entirely alone, thirty-six percent of them over the age of sixty-five. According to projections by the U.S. Census Bureau, by 2010 the number of people living alone will reach almost twenty-nine million—an increase of more than thirty percent since 1980—and a disproportionate share of these will be over sixty-five.[2]

As career patterns, housing patterns, mortality patterns, and social policies follow the lead of global capitalism, much of the world seems determined to adopt a lifestyle that will compound and reinforce the chronic sense of isolation that millions of individuals already feel, even when they are surrounded by well-meaning friends and family. The contradiction is that we have radically changed our environment, and yet our physiology has remained the same. However wealthy and technologically adept our societies have become, beneath the surface we are the same vulnerable creatures who huddled together against the terrors of thunderstorms sixty thousand years ago.

A Detour into Disconnection

The importance we assign to our place within a network of family and social relationships began to erode with the dawn of the industrial revolution. But even during the late Renaissance, long before men, women, and even children began to be forced out of their villages and into factories—long before anyone could even imagine corporate relocations, or crowds of solitary businesspeople waiting for the next flight to the next hub airport—the trend toward greater isolation was set in motion by a new cultural focus on the individual. That philosophical shift was reinforced by the rise of Protestant theology, which stressed individual responsibility, even in matters of salvation.

During the period immediately before industrialization, new generations resisted the idea of blindly following the old forms of authority—the Church fathers, classical figures such as Aristotle—and sought to reestablish first principles based on rational thought. During this Age of Reason, the English political philosopher Thomas Hobbes used the favored technique of the time, pure reason, to try to deduce the origins of the social and political structures that held the world together. He made the case that in a state of nature man was governed by nothing but his appetites and aversions and was free to do anything to get what he wanted, including murder his neighbor. In short, humanity lived in a benighted state of war. For humankind to escape from the homicidal pursuit of self-interest that was our natural inclination, social regulation had to be imposed from on high by a sovereign.

In *Leviathan*, his major treatise on the origins of governing powers, Hobbes examined the path humanity might have followed in order to arrive at what he christened the "social contract," an implicit agreement to behave decently toward one another rather than to do what comes naturally. Life in humanity's unregulated state of nature, Hobbes wrote, was "solitary, poor, nasty, brutish, and short."[3] What Hobbes assumed to be natural was very much in line with what later generations would describe—in a gross oversimplification—as Darwinian, by which they meant "nature red in tooth and claw."

Hobbes's limited assessment of human nature still informs (or misinforms) much political and economic discussion today. But Hobbes's pessimism was based not on empirical research but on assumptions seemingly derived from his own experience. He lived his entire life in a time when England was racked by religious strife and persecution, including a brutal civil war and the beheading of the king. Born in 1588, the year the Spanish Armada attempted to invade England, he later claimed, "Fear and I were twins."[4]

The world still provides ample evidence of civil war and religious persecution, but recent work in anthropology and evolutionary biology presents a picture of early human relations that is significantly at odds with what Hobbes called "the war of all against all." Which is not to say that early humans were "noble savages," any more than modern humans are entirely peaceloving and altruistic. Hobbes was not wrong in asserting that our ancestors could be brutish, but he was certainly off base when he described their existence as solitary. The greater error, however, was in assuming that their lives were unregulated.

No one can deny that competitiveness, envy, hatred, cruelty, and betrayal are aspects of human nature, and that these negatives are all well represented in human history. The point which the Hobbesian analysis misses is that, if such ruthlessness were, in fact, the defining essence of human nature, we never would have evolved our way out of the rain forest, much less the grasslands of eastern Africa. "Human nature" is a complex of many factors that range from the biological to the purely cultural. As we will explore more fully in later chapters, the driving force of our advance as a species has not been our tendency to be brutally self-interested, but our ability to be socially cooperative.

While Hobbes assumed that nature is an unregulated state, the primary task of every organism in nature is, in fact, to regulate itself in response to its environment. For social animals, a highly significant part of that environment is "each other," and thus members of families, tribes, and villages regulate themselves as individuals while also influencing one another through what we have called co-regulation. This system of checks and balances involves physiology as well as behavior. Co-regulation takes place, for example, not only when the presence of sexually receptive females increases the level of testosterone in the blood of nearby males, but when apes spend hours grooming each other. They spend ten percent of their time engaged in this activity, but cleaning fur is the least of it. The more important objective is to promote troop harmony and cohesion. And governing life over all, including human social behavior, is the ultimate, self-regulating, co-regulating process called evolution through natural selection.

In his *Autobiography*, first published in 1887, Charles Darwin provided a beautifully simple description of this regulatory process. He wrote that, over the course of generations, "favorable variations would tend to be preserved and unfavorable ones to be destroyed."[5] Even Darwin, however, wondered how this mechanism of diversification, pressure from the environment, and selection could account for certain socially generous behaviors, such as a bee that stings an intruder that threatens her hive, given that she disembowels herself in the process. Today we know that certain ants literally explode as weapons in defense of the colony. Others spend their lives as animate barricades, or as storage receptacles for food, hanging from the ceiling of a nest. How could such extreme devotion to social bonds and the social good be based on traits transmitted from one generation to the next, when the worker ant or bee does not, itself, reproduce? Darwin found this a major stumbling block, a seeming paradox that appeared "insuperable, and actually fatal to my whole theory."

Roughly a hundred years after Darwin's musings, and with the advantage of a thorough knowledge of genetics that did not exist in the great naturalist's day, another British biologist, William D. Hamilton, unraveled the evolutionary underpinnings of social bonds ranging from the self-sacrificing bee to the kindly human grandparents basking in the pleasure of their assembled offspring. Crucial to Hamilton's refinement of Darwin's basic theory was the realization that natural selection takes place not so much at the level of the individual or the group, but at the level of the gene.

Like Darwin, Hamilton could observe that a bird or a prairie dog that gives a warning call to save the group makes itself the one most likely to be carried off by the approaching hawk. One way that such "other-directed" behavior makes evolutionary sense is this: The prosocial gene or constellation of genes that drives an animal to sound the alarm, even at the cost of its own life, is shared by many of its closest relatives, including the selfless lookout's nieces or nephews. So even if the lookout dies young in the process, having lots of surviving nieces and nephews improves the rate of propagation of the genes that biased the lookout to do what it did. Over time, a characteristic that even modestly increases the survival and reproductive rate of individuals carrying the particular genes for a particular characteristic can spread until that characteristic becomes "species typical."

Hamilton's theory for how a gene for helping others, even at the cost of one's own life, could be passed along was first called "kin selection" and is now called "inclusive fitness." It led to a broader concept called "reciprocal altruism." Humans extend altruistic acts to people who are not their blood relatives. Such behavior is species typical because altruism reinforces social connection, and social connection, along with the genetic dread of loneliness that is its flip side, helped our ancestors survive.

In his book *Adaptation and Natural Selection* (1966), the evolutionary biologist George Williams summed up the idea with a direct contradiction of Hobbes's notion of early human existence as a constant state of battle: "Simply stated, an individual who maximizes his friendships and minimizes his antagonisms will have an evolutionary advantage, and selection should favor those characters that promote the optimization of personal relationships."[6] And what we know from studies of the few pre-industrial, pre-agricultural social groups left on the planet supports this observation.

Living on the Edge

In the Kalahari Desert of northwestern Botswana live tribes of hunter gatherers called the !Kung San. They are often described by outsiders as living proof of the survival advantages of strong social bonds. "Most creatures get what they need to live from their physical surrounding," Roy Baumeister wrote. "Humans, in contrast, get what they need from each other, and from their culture."[7] A quick look at the !Kung's physical environment shows us why they are so deeply embedded in each other's lives.[8]

Coming alone into the !Kung's home range, a city dweller would find miles and miles of dust and scrub vegetation. If dehydration didn't kill him first, that same city dweller would most likely starve to death pretty quickly. Yet archaeological excavations show that this region has been occupied by this same cultural group, living the same way in the same spot, for more than eleven thousand years. In the Kalahari, rainfall is scarce, summer temperatures exceed 110 degrees Fahrenheit, winter temperatures dip below freezing, and, given the presence of lions, "fast food" could easily refer to you or me. Living off the land in a place this harsh makes clear why early humans could ill afford to be nasty and brutish, at least not toward members of their own social group.

Even though the !Kung live in the midst of seemingly limitless real estate, a !Kung village is half a dozen huts tightly clustered around a small, cleared circle. Despite any desire for privacy, all doors face in toward the communal space. If you were to spend the night in such a village and see lions' eyes gleaming in the darkness just outside the ring of cooking fires, you might begin to appreciate why, for early humans, feelings of isolation were linked with fear, the fear that still remains at the core of our experience of loneliness.

Two anthropologists, Irven Devore and Richard Lee, first made contact with the !Kung living in the Gobe area of the Kalahari in 1963. Six years later a young woman named Marjorie Shostak arrived in Gobe for a two-year stay. She had no particular training in fieldwork—she was simply in Africa with her husband, the physician and anthropologist Mel Konner. But she decided to make use of her time by becoming fluent in the !Kung language and trying to get beyond the cultural and professional barriers to understand hunter-gatherer life on a personal level. The result was a book entitled *Nisa: The Life and Words of a !Kung Woman*, in which Shostak's account of life among the !Kung was interspersed with vivid monologues by the woman she called Nisa. The book became a literary sensation because it did not portray ancestral society as a war of all against all, or as a tableau vivant of the noble savage. Instead, it presented ancestral life as a soap opera, a tangle of intense social linkages in all their messy melodrama.

For months, Marjorie Shostak engaged in the !Kung San equivalent of sitting around the kitchen table with a cup of coffee. "Village life is so intimate," she concluded, "that a division between domestic and public life . . . is largely meaningless." The stories she compiled, stories independently corroborated by other fieldwork and dozens of interviews with other !Kung women, were filled with obscene jokes and lots of bed (or more accurately hut) hopping.

Even with the (admittedly brutish) rigors of avoiding hungry predators while finding enough to eat, it seemed that a vast amount of !Kung men and women's mental and emotional energy was devoted to managing social commitments. The opposite of solitary, life among the !Kung involves juggling relationships with a spouse and children, ever-present in-laws and other family members, assorted friends, enemies, and rivals who, nonetheless, contribute to one's survival, as well as a succession of lovers on the side.

The stories in *Nisa*, as well as the more straightforward accounts of traditional researchers, show that when !Kung women are not out gathering, or !Kung men off on a hunt, they spend a surprisingly large amount of time singing or composing songs, playing musical instruments, sewing intricate bead designs, telling stories, playing games, visiting, or just lying around chatting. They have no written language, but people sit together and talk for hours, repeating the same stories again and again. They have no calendar, but mark life as a progression of social events, from a baby's first social smiling, to first words, all the way to senescence and death.

This simple human society is a self-regulating system far more sophisticated than an ant colony or a beehive, but it operates on the same basic principle that each individual's actions are shaped and constrained by the actions of other individuals. Social insects co-regulate by way of chemical communication; humans, having far greater behavioral latitude, rely heavily on culture, but the fact that humans can teach and learn nongenetic (cultural) information about how to behave does not mean that they have left body chemistry behind.

The most significant way in which the !Kung demonstrate their predilection for closeness and co-regulation is in their approach to childrearing. Infants have access to the breast every moment of the day or night for at least the first three years of life. They nurse on demand several times an hour. They sleep by their mothers at night, and during the day are carried in a sling, skin to skin. Mothers carry their kids, on average, fifteen hundred miles a year. Separation, when it comes, is initiated by the child as soon as he or she wants to venture forth and play with other children. Even so, lastborn children will sometimes nurse until age five or even longer, when the ridicule of other youngsters—a natural form of social regulation—makes them stop. On average, then, !Kung children have forty-four months of close attention from, and body contact with, their mothers.

"Give me" is one of the first phrases a !Kung child learns, and the cultural norm demands generous and free exchange. In fact, !Kung life is so completely egalitarian—an almost universal finding among pre-agricultural societies living this close to the edge—that there is no chief or headman. All food is shared. Access to land is collective, and stinginess is a serious matter, punished by social exclusion. The most successful hunters must be self-deprecating. They carry arrows given to them by others, and the person whose arrow brings down the animal is considered the provider of the meat and oversees its distribution. They have gift-giving rituals, name-sharing rituals, and as the ultimate co-regulating social behavior, seasonal congregations to bring together separate bands and to engage in ecstatic "trance" dancing.

Make no mistake—the life of the !Kung is not "Eden in the outback," as some have dubbed it. Hemmed in by farming villages and limited to a depleted range, the !Kung today are not necessarily a perfect replica of hunter-gatherer life during all of human evolution. They are only one vestigial pocket, and no doubt their own customs have evolved over the past forty thousand years, even as the global environment has seen many changes. And their generally peaceful and cooperative social life can be punctuated by co-regulation that takes the form of violence. With an estimated twenty-two killings in five decades, the fifteen-hundred-member band studied by Mel Konner had a higher murder rate than the United States.[9]

Nonetheless, the !Kung's way of life is the best illustration we have of the social forces that shaped our human ancestors throughout their long evolutionary trek from small hominid ape called *Australopithecus afarensis* to a much smarter, and much more cooperative and even altruistic, species called *Homo sapiens.* And every preagricultural society we know about has this same basic structure. Against harsh odds they barely survive, but the fact that they survive at all they owe to the dense web of social contacts and the vast number of reciprocal commitments they maintain. In this state of nature, connection and social cooperation did not have to be imposed by a primitive form of the state, or by an English philosopher. Nature *is* connection. Which is why disconnection leads to such dysregulation and damage, not just at the level of society, but at the level of the cell.

Tit for Tat

Anthropology is necessarily an observational science, and no matter how carefully you observe primate societies or pastoral human groups, what you see does not constitute a controlled experiment that could demonstrate conclusively how these social structures actually evolved. Fortunately, computers allow us to complement observation with simulations. In the early 1980s a political scientist named Robert Axelrod came up with a computerized way of exploring the same question that Thomas Hobbes had tried to reason out: how social cooperation emerges, the kind we depend on not just for our political or economic well-being, but for proper physiological functioning as well.

Axelrod wanted to see if a sense of connection and social cooperation required an infusion of abstract moral reasoning, or top-down coercion, or if it could have developed as a natural phenomenon. To find out, he invited fourteen experts to submit computer programs designed to compete against others in finding the ideal solution to a well-known puzzle called the Prisoner's Dilemma. Axelrod unleashed these bits of computer code into a contest that was its own Darwinian cyber jungle, then sat back to see where their decisionmaking algorithms would lead.

In the Prisoner's Dilemma, two accomplices are arrested and interrogated in separate rooms. The authorities give each prisoner the same choice: Confess your shared guilt (in effect, betray your partner) or remain silent (and be loyal to your partner). If one betrays and the other stays silent, the defector goes free, and the silent, loyal one spends ten years in jail. If both remain loyal, both get six months. If both betray the other, both get six years.

The Prisoner's Dilemma is like a game show in which you can keep your toaster oven and your new washer-dryer or risk it all to see what's behind door number three. The choices are perplexing enough even without the kicker, which, is, of course, that neither suspect knows what the other is going to do.

Wrap this problem in scales, fur, or feathers, and you have the dilemma that was faced by our evolutionary ancestors all the way back to tree shrews and lizards. All of life represents a working out of the cost-benefit ratio of cooperation versus aggression. As the evolutionary biologist Martin Nowak suggests, "Perhaps the most remarkable aspect of evolution is its ability to generate cooperation in a competitive world."[10] We humans are at the top of the food chain because we are the species most adept at behaving generously while also accruing the benefits of competition.

Working in a far more comfortable university setting, where natural selection is limited to peer review and tenure decisions, Axelrod set up a round-robin tournament in which pairs of computer programs played the part of the prisoners. In the course of five games, each consisting of two hundred moves, each of the bits of code pursued the behavior it had been programmed to follow. These ranged from unwavering loyalty to reflexive betrayal. After this first round, Axelrod acted as a kind of Darwinian philosopher king. He tallied the players' "fitness" scores, determined by how little time each "prisoner" spent in jail, and declared the winner. He then circulated the results and called for another, follow-up tournament. This time he received sixty-two entrants from six countries, programs designed by computer hobbyists, evolutionary biologists, physicists, and computer scientists.

In this second round, as in the first, the ideal strategy for social behavior that emerged was a program called "Tit for Tat." Ironically, it was the simplest scheme of the bunch—only five lines of code. Its rules were as follows: On the first pairing with any other program, Tit for Tat would remain loyal. Thereafter, it would follow its competitor's example, doing whatever the other program did on each successive move. Loyalty would produce loyalty, in other words, and betrayal would produce betrayal. If Tit for Tat met with a potential ally, it would form an alliance, and both parties would benefit. Upon meeting up with a defector, Tit for Tat would cut its losses, refusing to exhibit any loyalty until the other guy reformed and stopped his betrayals. So while Tit for Tat led with a bias toward cooperative social connection and its benefits, it also avoided being taken for a ride. But by its constant receptivity to the idea of goodwill, it also avoided the destructive downward spiral of selfish and antisocial behavior.

Axelrod's experiment became a classic because it illustrates how, without any conscious awareness, organisms could evolve societies based on positive social interaction for no reason other than the strategy's superior long-term results. There was no moral consciousness in this generally benign and cooperative approach. Nature does not place any "value" on social solidarity any more than Tit for Tat does. A strong impulse in favor of connection simply produces better outcomes for survival. So, unless shown by immediate experience that loyalty is foolish, most social animals will look out for direct kin or close associates. But humanity is a special case. The story of our evolution, as opposed to the evolution of the other hominid apes, is the story of widening circles of, and an ever-increasing role for, social cooperation. With such behavior woven into our DNA through natural selection, the intensity of social bonds increased as well.

Which is why there is one more piece to the puzzle. A computer program like Tit for Tat does not become miserable or stressed to the point of dysregulation when it is betrayed or socially excluded. Humans do. Survival of the fittest led to creatures that were obligatorily gregarious. These were creatures that were deeply connected to one another through a complex web of physiological signals and sensations. These signals and sensations created links among outside stimuli, pleasure/pain, and behavior. The physiological sensations are called emotions, and it is their role to maintain the ever-stronger bonds that made humans human.

From Genes to Behavior

Hostility and social atomization ("the war of all against all") are not only associated with social chaos and with loneliness; they are associated with increased levels of chronic illness and early death. Add that to the extremely harsh conditions in which our ancient hunter-gatherer forebears lived, and we see that natural selection would have exacted a heavy price for any behavior that lessened the fitness of the group or the individual. The margins for error were small. As they worked their way out of the forest and onto the plains, early humans lived in an environment of evolutionary adaptation in which the most perplexing challenge no longer came from the flora and fauna surrounding them. Reading one another, sometimes deceiving one another, maintaining peace with one another despite the quirks of human sentiments and ever-increasing human intelligence—this was the next major arena in which natural selection would separate the most advantageous genes from ones that biased us toward less adaptive characteristics. In determining whether or not your genes would make it into the next generation, being slow to catch on to the social vibe became a more common threat than being mauled by a lion or bitten by a snake. And it was this need to "catch on," to sense what others were thinking and feeling, that gave the advantage to the ever more refined sensorimotor approach that involves the social emotions, as well as social cognition.[11]

Over millions of years, and with lasting implications for our health today, this selective pressure shaped the receptors and transmitters of emotional signals. It also shaped their integration throughout our physiology, including our immune response, which attempts to minimize the effects of intrusion or injury, and our endocrine system, which promotes the adaptive orchestration of bodily functions through a network of blood-borne hormones. At the same time, with our genes interacting with the environment by way of the behaviors they encouraged, the challenges of our social milieu continued to drive our cognitive development.

There always were, and no doubt always will be, lapses in self-regulation, lapses in trust, and random acts of social ineptitude, greed, treachery, and even murder. Nonetheless, the bigger brain capable of keeping track of myriad complex connections, along with the drive to avoid the pain of loneliness by maintaining those connections, continued to provide sufficient survival advantage that prosocial features became standard equipment in all but a handful of human beings. As aversion to loneliness and attachment to other humans became almost universal, what is called an "environmentally stable adaptation," the greater number of participants relying on alliances, loyalty, social cooperation, caring, and concern made it even more advantageous to play by those rules, at least within our inner circle. All the more reason that the feeling of being excluded became all the more terrifying and disruptive.

Thus, tens of thousands of years after the first human societies were formed, we find human beings bound together by kinship, friendship, and all manner of tribal groupings, ranging from Ache head-hunting bands in Paraguay, to Red Sox Nation, to online multiple-user dimensions, to fans of *Star Trek*, to the Church of England. And while each of us may treasure occasional moments alone and many of us can relish blissful solitude for long stretches, not one of those billions of people in those millions of groups ever wants to feel the depressing and disruptive pain of loneliness.

All the same, the Hobbesian idea of the rugged individual clawing his way out of the mire in ruthless competition persists, from Ayn Rand's "virtue of selfishness" to Milton Friedman's reverence for unfettered markets. The evolutionary biologist Richard Dawkins no doubt reinforced Hobbesian assumptions when, in 1976, he used *The Selfish Gene* as the title for a book that has since sold more than a million copies. By his use of the term, Dawkins underscored the fact that nature is indifferent to the survival of any individual organism. On a functional level, natural selection chooses winners and losers from among genes, not from among particular plants or animals. From a nucleic acid's perspective, each individual is little more than an experimental animal for testing which genes produce the traits with the greatest survival advantage within a given environment. Which helps us understand why the "winning" genes that spread throughout a population are not necessarily the ones that appear most advantageous for any given individual in any given moment. The gene is driving a larger process.

Fair enough, but what Dawkins's provocative title neglected to say is that, even though the individual is simply the vehicle through which "selfish" DNA makes more of itself, advancement beyond a certain level eventually required the gene to modify the selfishness of that vehicle. Reproduction remains a dead end unless one's offspring also survive long enough to reproduce. A sea turtle, given the relative simplicity of the rules it follows to stay alive, can make do by simply laying thousands of eggs on the beach, returning to the sea, and hoping for the best. But mammals produce far fewer offspring than sea turtles with each roll of the reproductive dice. This shift in strategy includes a shift toward requiring that mothers maintain reciprocal bonds with their offspring in order for the offspring to survive. Higher up the evolutionary ladder, as animals developed even more complex problem solving, more social learning, and less dependence on reflexes and fixed action patterns, the need for maternal investment in each individual offspring ratcheted up.

Primates generally produce only one infant at a time, so a much higher degree of caretaking investment in each individual became the adaptive norm on the evolutionary path to you and me. Even for primate males, the scattershot strategy of "keep moving, love 'em and leave 'em" (and leave the caretaking to the females), which had been optimal on lower rungs of the evolutionary ladder, eventually gave way to increasingly intimate social bonds and increased paternal investment. This was not because our primeval forefathers took parenting workshops, but because, especially in humans, making a greater commitment to social bonds and to nurturing the young increased the chances that one's offspring, or even one's nieces and nephews, would live long enough to have offspring of their own. Success in that next iteration of the birth-death cycle is the only way to ensure that one's own DNA continues to be passed along.

I Feel, Therefore I Am

Like any number of other characteristics, the genetic propensity for desiring social connection and the propensity for feeling social pain in its absence are transmitted through bits of genetic information in our cells, coded as instructions for making proteins. The expression of these genes is dependent on environmental circumstances, whether real or merely perceived. Some of the proteins take the form of the hormones that carry messages in the blood. These messages serve to integrate different organ systems and to coordinate behavioral responses. One of the hormones is epinephrine, which can flood us with the cluster of sensations we know as arousal. Another small protein—the hormone oxytocin—promotes breastfeeding, soothing calm, and close connection. Other genetically orchestrated proteins give rise to neurotransmitters such as serotonin, which can elevate our mood or send us into despair, depending on its concentration in the brain. The genes provide the chemical carrots and sticks that guide behavior, but they depend on the sensory systems to actually interact with the environment. Signals that the senses receive from the environment trigger changes in the concentration and flow of these hormones and neurotransmitters. These chemicals serve as internal messages to prompt specific behavior—and this is when the genetic instructions at long last appear as individual differences in levels of anxiety, or agreeableness, or sensitivity to feelings of social isolation.

Historically, individuals with behavioral dispositions less well adapted to the environment did not survive—or they survived only marginally, or they did not survive long enough to produce as many offspring as those who were better adapted. Individuals with better-adapted behavioral repertoires lived to produce more children, or at least more children who themselves lived long enough to reproduce, allowing the genes responsible for those better-adapted traits to be passed along more widely.

Among ancestral humans, bonding with the larger group became the norm, but for different reasons depending on gender. Bonding gave hunter-gatherer females a survival advantage: The group meant safety, but it also meant being able to share maternal duties while taking care of other necessary business. Even among wild savannah baboons in Africa, individual differences in the capacity to form close relationships with other females have a significant effect on the rates of offspring survival, a factor that persists independent of the mother's dominance, which group she belongs to, or any other factor in the environment.[12]

Among early human males, puny scavengers who relied on sharp sticks as weapons, bonding together to form alliances became the norm for its political advantages (and political dominance led to better mating opportunities), and also because it provided strength in numbers for safety. But the greatest advantage of social connection and coordination may have been in the acquisition of large amounts of concentrated protein. Lions are anatomically much more formidable than humans when it comes to aggression, and even they rely on highly coordinated teamwork to bring down prey larger than themselves.

But adaptive advantage breaks along gender lines in a more fundamental way. That is because, in addition to natural selection, there is a second, equally powerful force driving evolution, which is called "sexual selection." It consists of two complementary elements: competition among males and female choice.

Anywhere in the mammalian lineage, females must devote considerable time and caloric resources to bearing and nursing the young. Near the top of the evolutionary ladder, a female chimpanzee will forage with an infant on her back until it is five years old. Males in most species of mammals invest only a few seconds of reproductive energy, and thus they can do well by a strategy of having sex with any available female, whenever they can, and leaving the rest to chance. But given the female's immense contribution of reproductive time and effort, it makes no sense for her to jump into the nest with just anybody. When it comes time to mate, female chimps, like females of other species, look for the best possible return on their investment, that is, whatever will increase their odds of having offspring that will survive to reproduce. "Unusually fit fathers tend to have unusually fit offspring," George Williams notes, so "it is to the female's advantage to be able to pick the most fit male available."[13]

Even among the lower evolutionary orders, female discretion leads males to compete, advertising their fitness with flashy tail feathers, or big muscles, or—among bullfrogs—loud, sustained croaking, until the female makes her choice. Sometimes male competition includes offering gifts to the female. Given that reproduction in females consumes calories as well as time and attention, male courtship in many species involves a "nuptial offering" rich in nutrients. In dung-rolling beetles, the gift is a massive ball of elephant dung. Among hanging flies it is a dead insect. With the praying mantis, the "gift" is the male's head, chomped off and eaten by the female during intercourse. The evolutionary biologist Robert Trivers summed up the situation this way: "One can, in effect, treat the sexes as if they were different species, the opposite sex being a resource relevant to producing maximum surviving offspring."[14]

Among most species of birds, maternal investment means not only producing eggs but keeping the eggs warm to incubate them, then feeding the nestlings until they fledge. In those species female choice is not satisfied with a male who would strut his stuff, reproduce wildly, and leave the rest to fate. Unless the father brings food back to the mother or shares in nest-sitting when she goes out to forage, eggs left behind could become too cold to hatch or be scrambled by the first predator to come along. So males began to offer not just a one-time nuptial contribution but the promise of ongoing provisioning and protection of the young. Linked through chemically and culturally forged social bonds, the mating pair evolved along with the caretaking father. The apotheosis of male parental investment is the central plot device in *March of the Pen-*

guins, the hugely successful documentary in which we see male Emperor penguins standing for months in ridiculously cold Antarctic winters, with their eggs and later their hatchlings resting on their feet, tucked under the warmth of the paternal paunch.

Among humans, the same kind of ongoing partnership between mates, called the "pair bond," combined with high male investment in protecting offspring, contributed to a tipping point in reproductive success. Parental teamwork meant not only that increasing numbers of children might survive, but that these creatures could afford to be more developmentally and behaviorally complex. Greater behavioral latitude led to greater diversity, which led to innovation, which led to more rapid cultural learning.

But even before the appearance of pair bonding and male parenting, according to what we have already introduced as the social brain hypothesis, the intelligence of primates had already increased, driven in large part by the requirements of managing increasingly complex social structures. The "social brain" gave rise to the expanded cerebral cortex in humans because it gave an advantage to individuals who could learn by social observation; recognize the shifting status of friends and foes; anticipate and coordinate efforts between two or more individuals, eventually relying on language to communicate with, reason with, teach, and deceive others; orchestrate relationships, ranging from pair bonds and families to friends, bands, and coalitions; navigate complex hierarchies, adhere to social norms, and absorb cultural developments; subjugate self-interest to the interest of the pair bond or social group in exchange for the possibility of long-term benefits; recruit support for the sanctioning of individuals who violate group norms; and do all this across time frames that stretch from the distant past to multiple possible futures.[15] It is worth noting that each of these subtle mental abilities requires the executive control function of the frontal lobes—the function that succumbs most easily to the disturbing force of feeling socially isolated.

Along the route to *Homo sapiens*, other epoch-making innovations emerged—the ability to walk upright, an opposable thumb for grasping, a shoulder good for throwing—that allowed our ancestors to increase the range of their immediate concerns. These anatomical features provided for both perception and action at a distance, which put a further premium on being able to think, plan, and communicate. Greater intelligence meant physically larger brains, which meant larger heads on infants, which demanded a wider pelvis on the mothers giving birth. But upright posture favored a relatively narrow pelvis to facilitate walking.

To resolve these competing anatomical demands, natural selection favored human infants that came into the world before their brains were fully formed. Cranial capacity could be kept to a reasonable level before birth, but the trade-off would be that cognitive, emotional, and social development would have to continue during the first months—even years—of life. This meant that all human infants would be born, to some extent, "half baked" and therefore utterly helpless for an extended period of time. Chimp babies can at least cling to their mothers from day one, but not so human babies. This prolonged period of complete dependency created intense pressure on the mothers, who still had to avoid predators and continue to forage for food—in pre-agricultural societies it is the women's gathering of roots and berries that provides the tribe's most reliable source of calories—all while feeding and otherwise caring for their helpless child.

This placed an even greater premium on bonding and on parental investment. For males as well as females, those who felt compelled to bond with their offspring and take care of them, even if they themselves had to subsist on less and endure more hardships, left behind more surviving relatives who carried their "socially connected" genes. Assuming normal variation in the genetically biased need for social connection, an ancestral male from, say, a hundred thousand years ago, might have had a social thermostat set so low that he could hoard food for himself without feeling much in the way of shame, guilt, or pain. He could have gone off on a three-day hunt, found the place where the antelope play, and simply never come back. He might have been oblivious to the absence of his family, or to the knowledge that they might be starving. Inured to loneliness as a signal of distress, hunting only to feed himself, he might have been better nourished than those who carried food back to camp and contributed to the good of all. But if his children did not survive long enough to mature and reproduce and nurture their own young, neither did his genes. (If his tribe did not survive, his children also would be less likely to survive.) The older, more purely selfish genes persisted, but their influence in the population at large shrank by reproductive attrition. Individual success was now driven by the ability to transcend selfishness and act on behalf of others. The selfish gene had given rise to a social brain and a different kind of social animal.

CHAPTER FIVE

the universal and the particular

There's a joke about the Finnish couple in which the wife complains to her husband that he never expresses his affection. He responds with consternation: "I told you I loved you when I married you. Why must I tell you again?"

We live in an age that does not approve of cultural stereotypes, and yet I think many of us would agree that each nation has its own signature behaviors. The English queue up in orderly fashion at the drop of a hat; Italians, less so. Berliners obeyed "Keep Off the Grass" signs under machine-gun fire during the revolution of 1919; Romans look upon a red traffic light as more a suggestion than a command.

Evolution wove our strongly social impulses into the essence of who we are as a species, but natural selection is not the whole story. There are also individual and cultural variations. For social insects, the behaviors that make the hive, the ant hill, or the termite mound an extended organism are genetically determined. For human beings, while behavior is genetically constrained, it is also personalized by all our own sometimes maddening quirks and complexities.

The tandem influence of inheritance and individuality is why each of us experiences loneliness in a way that is unique, idiosyncratic, and grounded in the particulars of our life history and our own immediate situation. At the same time, however, loneliness also subsumes structural elements that are universal. Lying somewhere between the individual and the universal, there is also a role for distinctive cultural influences.

Culture—whether determined by a family, a town, an ethnic community, or national identity—plays a role in shaping what we aspire to in our relationships, and thus what will ultimately satisfy us. In Finland, cultural norms dictate that a person won't feel odd or left out if he is not married. In Italy it is quite the opposite. But in spite of the importance assigned to marriage within the culture, fewer Italians than people of other nationalities identified their spouse as the one from whom they would expect help in an emergency.[1]

Friendship is another domain influenced by national identity. Germans and Austrians report having the smallest number of friends, followed by the British and Italians, with Americans reporting the highest number.[2] Then again, it may be that Americans simply define the concept of friend in a broader and more casual way than do people from other cultures.

Conflicts between cultural norms and our own desires can further complicate and sometimes camouflage our experience of loneliness. Web culture may suggest that being able to list a thousand "friends" on my personal page is what I should want; a different culture suggests that knowing everybody at the trade show and getting invited to the best hospitality suite with the open bar and the huge cocktail shrimp should be a major objective. Our media culture seems to have convinced millions that becoming "famous" via YouTube or reality TV, even if it involves personal humiliation, will make them happy. But then, all too often, people who have done everything right according to the cultural dictates they accept can still be left asking "Why am I so miserable?" They may be unable to articulate, or even to entertain, the thought that, despite their culturally endorsed achievements, they lack the meaningful connections that would assuage their sense of personal isolation.

A Man Apart

One gentleman from our study of older Chicago residents, Mr. Diamantides, seems like a poster child for the power of positive thinking as well as a certain kind of social savvy. When you ask how he is, his response is an emphatic "I'm wonderful. How are *you*?" Dapper and energetic, a sharp dresser, Mr. Diamantides has worked in retail all his life. "I connect well with people," he says. "It's easy for me—I'm Greek!" He even completed a year and a half of college studying psychology, "just so I could understand people." When he talks about social connection, he peppers his description of his life with phrases like "I'm just lucky"; "I'm blessed"; "Attitude is everything." He is also proud to say that he knows a great many movers and shakers. "I have a wealthy clientele . . . but my customers treat me well because I really like them. I make people feel important. I make them feel special."

In childhood Mr. Diamantides suffered no particular traumas, though the stigma of being "from the wrong side of the tracks" and maybe "not quite legit" stayed with him. "We were a little shady," he explains. "In those days, if my cousin showed up with some hot watches, you'd say 'Let me take a look,' no questions asked." But his parents, as he put it, were "really good to me."

Mr. Diamantides says that he has maintained his religious faith, but he does not attend church regularly. He was married once, briefly, but for more than twenty years he has lived alone. He has no children, but he has a large extended family—lots of cousins, nieces, and nephews: "In the family, even if you're wrong, you're right. It's a tremendous support system." All the same, he says he enjoys his solitude. "I'm in a people business, so when I come home, I'm thrilled just to be able to do what I want."

When asked to describe his loneliest moment, he mentioned the time when he was in his forties and both parents died: "I felt like an orphan." But when asked to describe his warmest moment of social connection, he was stumped: "I've had too many . . . it's hard to choose one." When pressed, he mentioned that once a longtime customer left him a thousand dollars in his will, and that a neighbor he hardly knew left him ten thousand, when all he had done was to take the neighbor to the doctor once or twice. Then Mr. Diamantides remembered the real emotional high. Some years back he and another man had gone in together to buy one "founder's" share in a start-up—they each put up ten thousand dollars—but there was nothing in writing. Years went by. They lost touch with each other. And then Mr. Diamantides received a check in the mail for five thousand—a long-awaited return on the investment. "My cousin told me, you know, there's nothing on paper . . . keep it! But no way. I got on the phone and I tracked this guy down. Took me weeks. I even had to call California. He nearly died when he heard it was me. And we split the money! I felt like a million. He was so surprised. I floated for days. It was the best feeling I ever had."

In conversation, Mr. Diamantides is so convincing in his claim that everything is great in his life that it is easy to assume he is "low in loneliness." You might peg him as an interesting anomaly, a man with no immediate family, no close friends, who doesn't "get out much," who nonetheless feels immensely satisfied with his social world. But when we looked below the surface we found quite a different story. Mr. Diamantides completed the UCLA Loneliness Scale for us. He also allowed us to test his sleep quality, blood pressure, morning levels of the stress hormone cortisol, and other factors. What the psychological test showed, and what the physiological data confirmed, was that Mr. Diamantides had one of the highest loneliness scores of all the people we had ever studied.

The clues to this apparent contradiction are scattered throughout his self-report. For instance, it is hard to discount all the people with whom Mr. Diamantides had fallen out. The cousin who said the wrong thing, the brother with whom he argued about money. "I can't forgive and forget," he told us. "I'm not hostile or bitter . . . you're just never in my heart again the same way." It turns out that Mr. Diamantides had been conned in certain financial dealings by his wife, and he decided that he would never allow himself to be so vulnerable again, so he essentially closed himself off to other people. Unfortunately, he has remained in that same emotional isolation for years.

For all the large number of individuals that Mr. Diamantides sees during the day, there is no one that he considers a friend. Not even within the closely knit extended family that was such a "great support system"—he rarely if ever speaks to or sees any of his relatives. And as for his greatest experiences of warmth and connection, they all involve money.

The point is that people can misuse their powers of cognition in their attempts to self-regulate the pain of feeling like an outsider. They can create a false persona—a practice commonly known as self-deception—that frames their life the way they want it to appear. By working very hard at it, sometimes they can convince themselves that "It's so because I say it's so." But the physiological and psychological effects of loneliness take their toll nevertheless.

Aspects of the Self

The role of subjective meaning in our sense of social connection is not all that different from the role of individualized, personal meaning in other aspects of our lives. You could have a designer come in and fill your bedroom, office, or den with expensive mementos, trophies, plaques, and photos apparently inscribed just for you by Elvis Presley and Vladimir Putin. Your visitors might be very impressed. But if all that stuff came from a prop shop, most likely when you walk into that room it still would not feel entirely right. You might be able to muster up some momentary ego gratification, but there would be no enduring sense of warmth and satisfaction, because those mementos and trophies would have no real meaning. In the same way, you can have all the "right" friends in terms of social prestige, in-group cachet, or business connections, or a spouse who is rich, brilliant, and fabulous looking, but if there is no deep, emotional resonance—specifically for you—then none of these relationships will satisfy the hunger for connection or ease the pain of feeling isolated.

Of course, in our daily experience, we don't think about cultural constraints on our subjective experience of isolation any more than we think about its formal structure. Whether loneliness has two dimensions or twelve is the kind of thing only psychological scientists worry about. Nevertheless, knowing the universal structure of the experience can be useful, especially if we are trying to do some renovations.

If I ask you to imagine a room, you are likely to come up with a certain memory, a certain color, a smell, a view out the window, or perhaps the furniture or the pictures on the wall. But it is also true that, when we are objective, quantitative, and attentive to what is common about any room, we recognize that any room we can imagine will have three fundamental dimensions: length, width, and height. You experience the room as one big rush of sensations—a *Gestalt*—but these three facets contribute to and constrain your experience of it. If you want to try to redesign this room to make it more pleasant or functional, you will have to take these three fundamental dimensions into account.

In the same way, if we want to make ourselves happier and healthier by enhancing our social satisfaction, it pays to understand the universals, one of which is "the self" itself.

The psychologists Wendi Gardner and Marilynn Brewer did a study to examine the ways in which people might describe themselves when asked the question "Who are you?"[3] They determined that self-descriptions can be categorized into three basic clusters:

1. *A personal, or intimate, self.* This is the "you" of your individual characteristics, without reference to anyone else. This dimension includes your height and weight, intelligence, athletic or musical ability, taste in music and literature, and other personal preferences, such as liking Tabasco over tapioca.
2. *A social or relational self.* This is who you are in relation to the people closest to you—your spouse, kids, friends, and neighbors. When you go to the PTA meeting you are little Zach's mom or dad. When you go to your spouse's office party, you are "the spouse of . . ." This is the part of you that would not exist without the other people in your life.
3. *A collective self.* This is the you that is the member of a certain ethnic group, has a certain national identity, belongs to certain pro-

> fessional or other associations, and roots for certain sports teams. Similar to the relational self, this part of the self would not exist without other people. What makes this self distinct is that these are broader social identities, linked to larger social groups, that may be less obviously a part of your day-to-day experience.

People see themselves in these three dimensions because these are the same three basic spheres within which humans have always operated. From our earliest evolutionary ancestry, human beings have been unique individuals with specific physical characteristics, personality traits, and likes and dislikes, but we've also always shared close bonds with mates and offspring, and we've always lived in larger social groupings, from extended families to tribes to nation states. The "self" behaves a little differently in each setting. When you define yourself as part of a group (the collective self), for instance, you may be more inclined to agree with other group members, even on beliefs that may seem irrational ("Of course the Cubs will win the World Series this year!"), than when you are thinking of yourself as a unique individual.

Brewer and Gardner demonstrated exactly this effect by priming college students to think of themselves in a collective context—namely as members of their particular college community—then measuring how long it took the students to agree or disagree with something another student from their college said. As expected, those students who had undergone this priming were faster to agree, and slower to disagree, with their group members than were those who had not been set up to think of themselves as part of a group.

When you think of your self at the level of your unique personal identity, it is only human to compare yourself with others and feel a twinge of hurt or jealousy if someone outperforms you at something important. When the person who bests you is a friend or family member, the defeat can be even more painful than when you lose to a stranger.[4] However, when your focus is on your family or community identity, it is easier to celebrate the triumphs of someone close to you as if these victories were your own. When Serena Williams defines herself as Venus Williams's sister, it makes it easier for Serena to enjoy a Venus championship. With the focus on family identity or family pride, each of these highly competitive tennis stars becomes part of the same unit, and as such, one sister's successes can be success for the other as well.

Three Degrees of Connection

In our research group, we compiled a vast amount of survey data documenting the structure underlying the ways people think about their connections to others. We subjected the data to factor analysis, a statistical sorting technique designed to uncover simple patterns in the relationships among variables. If you used factor analysis to analyze the features of a thousand rooms, the statistics would cluster to show that the essential factors that make a room a room are length, width, and height. Every room has them; there is no room without them. Other qualities such as "tacky" or "stuffy" or "green" would appear as variables standing outside the essential dimensions, one-offs that don't say anything universal about the nature of rooms.

By using this same quantitative sorting technique we found that the universal structure of loneliness aligned very nicely with the three dimensions of Brewer and Gardner's three-part construct of the self. For the self, the essential dimensions are personal, relational, and collective, onto which we can map the three corresponding categories of social connection: intimate connectedness, relational connectedness, and collective connectedness.[5] Humans have a need to be affirmed up close and personal, we have a need for a wider circle of friends and family, and we have a need to feel that we belong to certain collectives, whether it is the University of Michigan alumni association, the Welsh Fusiliers, the plumbers' union, or the Low Riders Motorcycle Club.

Not surprisingly, the three dimensions of the universal structure of loneliness are highly correlated. If you are happy in one (marriage, say), you tend to be happy in the others. Until, perhaps, perturbations in your environment throw you for a loop. Your husband suddenly dies, or you move to a new and alien community. A bereaved wife may have great friends, and these friends may do everything they can for her, but most often their support does not completely remove the deep pain of the loss of a life partner. When events knock one of the three legs of the stool out from under you—intimate, relational, or collective—the safe and comforting feeling of stability falls away, and even someone who has always felt intensely connected can begin to feel lonely.

However, we have also found that there is no absolute, one-to-one correlation between any of these objective, environmental indicators of social isolation and subjective experience. Marital status is one of the best predictors of intimate connectedness—that is, married people tend to be less lonely than single people—but not everyone finds marriage to be self-affirming. The nun, the explorer, the artist, or the hard-driving executive who does not marry may find meaning elsewhere. And we all know that close family connections can be a mixed blessing. The same is true for people who have more friends than they can keep up with. Believe it or not, for some people, the phone constantly ringing with invitations to fabulous soirees can become a source of stress. And while some of us are joiners, others are very private and need very little in the way of connection through group membership. On each of the three levels, the issue is not the quantity but the quality of relationships, as determined by our own subjective needs and preferences.

A former professor who described herself as "not a joiner" told me that she never appreciated her need for collective connection until she retired. It was only when she went back to live on her family's farm in the Midwest that she realized just how much being a part of her scholarly department and her prestigious university had meant to her. But once she had gone back home, she found new ways of filling the need:

> I belong to a very different group of people out here, people whose roots go back to pioneer days and who are steeped in the

> history of the area. Out here I don't have real friends yet (though the family connections are rewarding), but that wider kind of connection helps keep me from feeling too lonely, partly because it's just plain comfortable to feel like an insider, someone who belongs.

Similarly, many of us tend to ignore the collective aspect of social connection much of the time, then find ourselves surprisingly caught up in a group identity when a national emergency occurs, or when there is some insult to the dignity of a class of persons with which we identify. The attacks on New York and Washington on September 11, 2001, aroused the collective identity of Americans, just as the caricatures of Mohammed published in Denmark aroused the collective identity of even many Westernized Muslims. One person may watch a parade for immigrant rights and feel great: Look how diverse we are, yet we are all one city! Another person may watch the same event and feel threatened: This is not my town anymore . . . who *are* these people? We make meaning of such events—beautiful diversity, cheap labor, the end of the world as we know it—depending on many other factors in our lives and attitudes. And just as each of us represents the idiosyncratic within the universal, nothing says that your or my "idiosyncratic" is always going to be the same throughout our lifetime.

Over the past four decades, research by the psychologist Walter Mischel has demonstrated that, contrary to the idea of genetic determinism, people do not behave according to rigidly fixed traits that manifest themselves consistently across all situations.[6] It is not that there is no consistency, but that the consistency is situational and temporal. You may feel lonely every time you enter into a certain situation (the lunchroom in high school), even if, at the same period in your life, you feel socially satisfied very consistently in another context (band camp). Your susceptibility to loneliness may remain stable across time, but the situations that cause you to feel most acutely lonely in childhood or adolescence will most likely be different from the situations that induce acute loneliness when you are a young parent or an older adult.

Loneliness and Depression

An even greater challenge to sorting out the exact dimensions of loneliness is that it rarely travels alone. Much of the early research in psychology and psychiatry was conducted in clinical settings with individuals who were suffering from a number of maladies, often severe. The most common pairing was intense manifestations of both loneliness and depression.[7] Perhaps not surprisingly, then, the two constructs—loneliness and depression—were often lumped together.[8] "I feel lonely," for example, is a question on the Center for Epidemiologic Studies Depression Scale.[9]

Nonetheless, factor analysis tells us that loneliness and depression are, in fact, two distinct dimensions of experience.[10] Diagnostically, too, we know that depression is different, in part because it does not trigger the same constellation of responses that loneliness does. Loneliness prompts a desire to affiliate, but it also triggers feelings of threat and dread. As the experience grows more intense, the feeling of threat prompts a tendency to be critical of others. Loneliness reflects how you feel about your relationships. Depression reflects how you feel, period.

Although both are aversive, uncomfortable states, loneliness and depression are in many ways opposites. Loneliness, like hunger, is a warning to do something to alter an uncomfortable and possibly dangerous condition. Depression makes us apathetic. Whereas loneliness urges us to move forward, depression holds us back. But where depression and loneliness converge is in a diminished sense of personal control, which leads to passive coping. This induced passivity is one of the reasons that, despite the pain and urgency that loneliness imposes, it does not always lead to effective action. Loss of executive control leads to lack of persistence, and frustration leads to what the psychologist Martin Seligman has termed "learned helplessness."

Within the struggle to self-regulate, loneliness and depression are at their core a closely linked push and pull. The most primitive organisms operate entirely on the basis of paired opposites, essentially a gear for forward and another for reverse. This facilitates a simple, two-part decision—approach or withdraw—repeated endlessly as the organisms confront every stimulus. They approach to eat or to mate, and they withdraw to avoid negative sensations, which usually mean danger. Biological systems all the way up to and including human beings continue to operate on the basis of similar pairings.

Given the evidence that loneliness is an alarm signal, embedded in the genes, and that it serves a survival function, there may well be an equally adaptive social role for its opposite, depression. Imagine one of our long-ago ancestors, a young man in a hunter-gatherer community on the plains of Africa. Motivated by a feeling of social isolation, he makes an approach—he tries to court a woman, or he tries to join a hunting party or a political alliance—but, for whatever reason, he is rebuffed. Merely persisting and blundering ahead would be a waste of energy, most likely counterproductive, and maybe even dangerous. During this initial period of rejection, a mildly depressed mood (as well as the lack of persistence associated with loneliness) might be useful. By tempering the impulse to approach and affiliate, depressive feelings might encourage our awkward ancestor, his executive control now diminished by his sense of social exclusion, to back off long enough to analyze his situation: "Maybe I came on too strong?" "Maybe I should offer a gift to soften them up."[11] At the same time, the passivity of the depressed mood (and the passive coping that loneliness ultimately induces) would conserve his energy and resources.[12] Within a social hierarchy, when we have attempted an advance and failed, it can be to our advantage not merely to step back and rethink but to signal submissiveness.[13] In that delicate context, depressed affect could serve as the human equivalent of a dog rolling over on his back and showing his vulnerable belly. The real pain of depressive feelings might also be a means of social manipulation—a cue, similar to crying, that says "I need help" and solicits attention and care from those around us.[14] All in all, this inducement to lie low and to signal to others that we are not a threat might serve to minimize risk in social interactions during a time when we perceive that our social value is low, especially in relation to the intensity of our social wants.[15]

This kind of "go forward/back up" system may have worked long ago without so many of the negative consequences that we see today. In a world less socially complicated, with perhaps less mental anguish than modern humans now generate, the playing out of the sequence "approach/blunder/withdraw" followed by "regroup/resume normal activity" most likely occurred within a fairly brief time and without the need for the same cognitive sophistication required in today's complex social environment. Extrapolating from primate behavior, we can reasonably assume that social conflicts, like most threats during the early development of our species, led to fairly quick resolution—for good or ill. The limited cognitive powers of the earliest hunter-gatherers, and the harshness of their environment, would not have allowed them the luxury of long bouts of passive melancholy, ambivalence, and soul searching. Over many millennia, however, with increasing intellectual and psychosocial complexity, a simple sequence of "go/stop/go again" has evolved into a vicious cycle of ambivalence, isolation, and paralysis by analysis—the standoff in which loneliness and depressive feelings lock into a negative feedback loop, each intensifying the effects and the persistence of the other.

This is the situation in which we left our friend from Chapter One, Katie Bishop, sitting in front of the television, eating ice cream directly from the carton. If she were a character in a date movie, she might run down to Starbucks the next morning and spill her latte on the perfect someone, finding romance, companionship, and a wide social network of zany new friends. But then again, in real life, she might feel so low that she simply pulls the pillow over her head and hides under the covers until noon.

When we begin to look for the specific physiological pathways that lead from social isolation to increased illness and shortened life expectancy, we have to consider the possibility that loneliness and depression are both manifestations of some other overarching problem. We also have to take into account a long list of other variables that might show up in the same kinds of circumstances, any one of which might account for the effects we see. How can we determine whether it is actually loneliness, or one of these associated factors, that is driving the plot as the story unfolds?

There are three standard ways to identify associations and investigate causal relationships. The first is a cross-sectional study: You cast a wide net to gather many different types of people, and then you measure a variety of variables at a single point in time. The second is a longitudinal study, which means identifying a certain population and then following its members over a long period, making repeated measurements of certain variables as their lives play out from day to day. The third is random assignment and experimental manipulation.

Both cross-sectional and longitudinal studies can provide a wealth of useful data. The longitudinal approach also controls for a number of additional factors that cannot be dealt with satisfactorily in a cross-sectional study. For instance, whether an adult had a secure or insecure attachment with his mother in infancy may not be something that can be measured reliably now. However, each participant in a longitudinal study serves as her own control—the study follows the same person, after all, and her past remains the same. In a longitudinal study, then, in which the focus is on changes in loneliness and related variables over time, we separate and evaluate the effects of these changes from those that do not change across time, such as infant attachment style. Still, neither of these approaches can tell us definitely that we have found direct cause and effect. Even if we can demonstrate a strong association between loneliness and certain other factors in a longitudinal design, and even if we have ruled out all known alternative accounts, it does not mean that we have shown convincingly enough to overcome the skepticism of good science that one factor *causes* another. This is when experimental manipulation becomes particularly useful. To sort out the constellation of variables surrounding loneliness, and to determine what is the most likely cause of what, my colleagues and I used all three approaches.

Manipulating the Mind

For our cross-sectional analysis, we went back to the large population of Ohio State students that had supplied volunteers for our dichotic listening test. We refined our sample down to 135 participants, 44 of them high in loneliness, 46 average, and 45 low in loneliness, with each subset equally divided between men and women.[16] During one day and night at the General Clinical Research Center of the OSU Hospital, we subjected these students to such a wide array of psychological tests that we might have been packing them off for a mission to Mars. This allowed us to develop a precise statistical profile of other personality factors as they appear in association with varying degrees of loneliness. In other words, this study population gave us a clear picture of the full psychological drama accompanying loneliness as it occurs in the day-to-day lives of a great many people observed during a specific period of time. The cluster of characteristics we found were the ones we had anticipated: depressed affect, shyness, low self-esteem, anxiety, hostility, pessimism, low agreeableness, neuroticism, introversion, and fear of negative evaluation.[17]

Given the complexity of that implicit drama, the next challenge was to see if we could demonstrate through a controlled experiment that loneliness played a leading rather than just a supporting role. A controlled experiment means studying people in a situation in which you can hold certain variables constant while you manipulate some other variable. Such an experiment also requires that the participants be randomly assigned to different levels of the manipulation taking place.

To manipulate levels of perceived loneliness, we enlisted David Spiegel, a psychiatrist at Stanford University, to hypnotize our experimental subjects. Using precisely worded scripts, we guided our hypnotized student volunteers to re-experience moments in their lives that summoned up either profound feelings of loneliness or profound feelings of social connectedness. With some individuals we induced loneliness in their first hypnotic state and social connectedness in their second; with the others the order was reversed. Before and after each hypnosis we administered the revised UCLA Loneliness Scale to ensure that the hypnosis had induced the desired emotional state.[18]

Earlier, Spiegel had done a classic experiment with Harvard's Steven Kosslyn to demonstrate that hypnotic suggestion was not merely an extreme case of suggestion, coercion, and compliance. This earlier study focused on color perception: Hypnotized participants would be shown an image while they were told either that it was in color or that it was in black-and-white; the hypnotic suggestion sometimes matched the actual image and sometimes did not. PET scans administered during the hypnosis showed that the subjects' brains were physically registering color or black-and-white according to the hypnotic suggestion, even when it was contrary to fact. In terms of the brain's response, then, the induced experience was as real as real can get.[19]

With each of the Stanford students, after the hypnotic induction had produced feelings of loneliness or social connectedness, we administered the same psychological tests that we had administered earlier to our OSU students. As displayed in Figure 5, the results were a match. Examining the graphs side by side was like watching *CSI* when the forensic experts match up fingerprints.

In the top panel of the figure, the two jagged lines compare the test results for the OSU students in the top twenty percent in terms of loneliness (the solid line) and the OSU students in the bottom twenty percent in terms of loneliness (the dashed line). The students high in loneliness, compared to those low in loneliness, reported lower levels of social support, higher levels of shyness, poorer social skills, higher anger, higher anxiety, lower self-esteem, higher fear of negative evaluation, lower optimism, lower positive mood, and higher negative mood.

In the bottom panel, the two jagged lines compare the test results for individual Stanford students when they had been hypnotically induced to feel high in loneliness, and when these same individual students were induced to feel low in loneliness. Their test results for the eleven other characteristics being measured—mood, optimism, social skills, and so on—followed very much the same pattern.

Merely by manipulating feelings of loneliness and social connectedness, we had managed to produce corresponding appearances by, with corresponding levels of intensity from, all the other players in the drama. Loneliness, then, definitely had a starring role.

Moreover, we had demonstrated yet again that lonely individuals are not a breed apart. Any of us can succumb to loneliness, and along with it, all the other characteristics that travel as its entourage.

The surveys of the Ohio State students as well as the manipulations by way of hypnosis showed the effect of loneliness on thoughts, moods, self-regulation, even personal characteristics such as shyness and self-esteem—in the moment. But what about chronic loneliness? Leaving participants in an unpleasant and unhealthful state over time would be exceedingly unethical, so we could not induce persistent feelings of social isolation through manipulation. Longitudinal research is an ethical alternative, which is why we initiated our longitudinal study of middle-aged and older adults from the greater Chicago area.

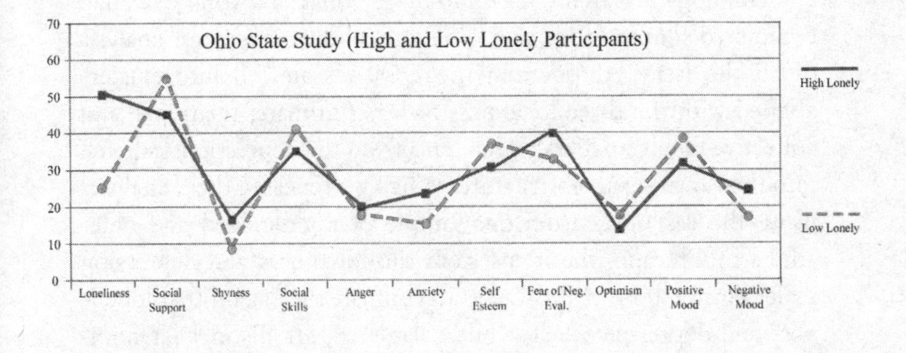

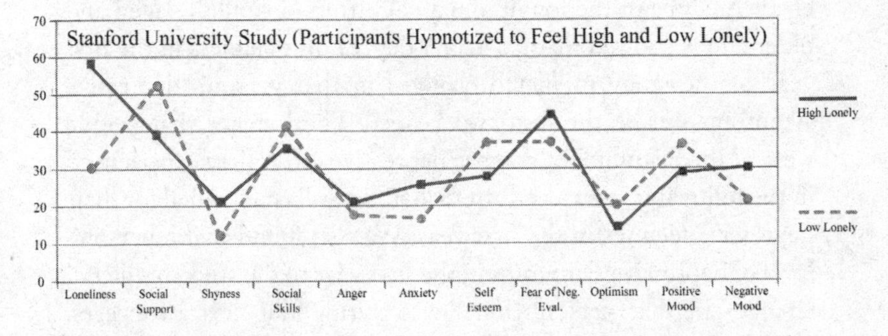

FIGURE 5. *Top panel:* comparison of characteristics of very lonely individuals with those of not at all lonely individuals. *Bottom panel:* comparison of characteristics of individuals induced to feel lonely with those of the same individuals induced to feel nonlonely.

Restoring the Whole

To accomplish a precise measurement of the effects associated with chronic loneliness and changes in loneliness over time, we identified a subset of individuals from our Chicago study population who were truly a representative sample, the kind that news organizations use in order to predict elections on the basis of survey data. We used a quota sampling strategy at both the household and individual levels to achieve an approximately equal distribution of participants of African-American, Latino, and other European ancestry, as well as an equal number of men and women in each group, all between the ages of fifty and sixty-seven years.

We asked each participant to complete the revised UCLA Loneliness Scale as well as a measure of depression commonly used in epidemiology research. Our Ohio State University volunteers had completed similar scales, and when we conducted a factor analysis of all the items, those from the UCLA scale fell into clusters aligned with the three loneliness factors (intimate, relational, and collective connectedness). The items from the depression scale fell into their own separate structure. When we repeated these analyses using the responses from our sample of middle-aged and older adults in Chicago, the items from the loneliness and depression scales clustered in such a way as to confirm once again that loneliness and depressed affect, while correlated, are distinct phenomena.[20] Analysis of the longitudinal data from our middle-aged and older adults showed that a person's degree of loneliness in the first year of the study predicted changes in that person's depressive symptoms during the next two years.[21] The lonelier that people were at the beginning, the more depressive affect they experienced in the following years, even after we statistically controlled for their depressive feelings in the first year. We also found that a person's level of depressive symptoms in the first year of the study predicted changes in that person's loneliness during the next two years. Those who felt depressed withdrew from others and became lonelier over time. So here too was the stop-and-go mechanism of loneliness and depressive symptoms we had postulated, working in opposition to create a pernicious cycle of learned helplessness and passive coping.

Most important, these studies probing cause and effect suggested a way to get beyond the Catch-22 embedded in our experience of isolation. If the self-defeating symptoms of loneliness can be externally induced by hypnotic manipulation of memories and feelings, and if they can change over time as a result of real—but also perceived—changes in one's social environment, then with increased awareness and effort, there should be a way for lonely people to learn to manipulate those same perceptions, cognitions, and emotions themselves.

But before we examine that possibility, there is one more mystery to pin down. All the evidence points to loneliness as a ringleader that brought at least eleven other, associated emotional states to the scene of the crime, that crime being a sometimes life-threatening assault on physical health as well as emotional well-being. So loneliness was on the scene and exerting a lot of influence—but how can we be sure that loneliness itself was the actual "perp"? How can we be sure that the serious declines in health and well-being we observed over time were triggered by something so intangible as a subjective sense of social isolation? If loneliness has the power to actually cause illness, what is its modus operandi?

CHAPTER SIX

the wear and tear of loneliness

In *The Triple Helix*, the evolutionary geneticist Richard Lewontin described organisms as "electro-mechanical devices" that, for purely thermodynamic reasons, succumb to wear and tear. This wear and tear, Lewontin tells us, contributes to a general decline in function and, eventually, to death. Or, as the novelist John Irving puts it, "We are all terminal cases."[1]

Most of us do what we can to resist our inevitable decline, but we tend to think of staying healthy in terms of avoiding disease and injury. In 1948, however, the World Health Organization defined health as "not merely the absence of infirmity" but as a state of "complete physical, mental, and social well-being."[2] Even so, we most commonly measure our health as a readout of physical findings and test results from our last visit to the doctor.

The study of loneliness expands our focus to include social and emotional influences that don't show up immediately on an X-ray or in a blood test and yet can have an enormous impact over time.

In the early 1990s, when I was at Ohio State University studying social influences on physiological processes and health, I was asked to join the MacArthur Foundation Network on Mind-Body Interactions. This group of neuroscientists, endocrinologists, immunologists, psychiatrists, psychologists, sleep researchers, and others had formed a joint task force to investigate the mind-body problem—the question of how our mental life and our biology are interrelated.

The influence of social isolation on health seemed an ideal problem for us to tackle. A dozen years earlier, the epidemiologist Lisa Berkman had found that men and women with few ties to others were two to three times more likely to die in a nine-year follow-up period than those who had many more contacts. People with few social ties were at increased risk of dying from ischemic heart disease, cerebrovascular and circulatory disease, cancer, and a broader category that included respiratory, gastrointestinal, and all other causes of death.[3]

In 1988 an article in *Science* reviewed subsequent research, and that meta-analysis indicated that social isolation is on a par with high blood pressure, obesity, lack of exercise, or smoking as a risk factor for illness and early death.[4] For some time the most common explanation for this sizeable effect has been the "social control hypothesis." This theory holds that, in the absence of a spouse or close friends who might provide material help or a more positive influence, individuals may have a greater tendency to gain weight, to drink too much, or to skip exercise. The thought is that this neglect may account for the effects on health that show up in studies of isolation.

The notion sounds plausible. Not too long ago, both in Paris and in Chicago, older residents died during heat waves largely because they were cut off from other people who might have helped them adapt to the temporary extremes. In Paris, especially, the number of deaths was inordinately high because the heat wave occurred in August when families were away on holiday, and when many older relatives were left alone to fend for themselves.

But epidemiological research done on the heels of the analysis published in *Science* determined that the health effect associated with isolation was statistically too large and too dramatic to be attributed entirely to differences in behavior. The psychologist Dan Russell and his colleagues later confirmed the limits of the social control hypothesis when they examined the health histories of 3,097 people aged sixty-five years or older residing in two rural counties in

Iowa.[5] The individuals with the highest baseline scores for loneliness were also the ones most likely to be admitted to a nursing home over a four-year period. Furthermore, their degree of objective social support—whether or not they had a niece who came by to help out, or a neighbor who would drive them to the clinic—was not a significant predictor of the need for increased care once loneliness was taken into account.

We had a hunch that what mattered was not the number of social interactions, nor the degree to which other people provided practical benefit, but the degree to which social interactions satisfied an individual's specific, subjective need for connection. Earlier research in which participants were asked to fill in a diary at certain moments of each day had shown that the amount of time spent with others and the frequency of interaction did not add much to the prediction of loneliness. What did predict loneliness was, again, an issue of quality: the individuals' ratings of the meaningfulness, or the meaninglessness, of their encounters with other people.[6] But saying that lack of meaning in one's social encounters could become as injurious as obesity, lack of exercise, or the inhalation of carcinogenic smoke still sounded like a stretch.

To test our hypothesis that subjective satisfaction played a major role, we turned once again to our student volunteers from Ohio State. In addition to the battery of psychological tests we had already administered to these very patient kids, we measured cardiovascular functioning and drew blood at various times of day for endocrine and immune tests. We also kept these volunteers overnight at the University Hospital to measure certain aspects of their sleep.[7] Even after they went home, we continued to evaluate their sleep during five consecutive nights. To help keep track of their daytime perceptions as well as their physical reactions, we gave them beepers, which they carried for seven days. Nine times a day we beeped them, and each time the beeper went off they sat down and completed a short questionnaire about their moods, activities, thoughts, and feelings—the kind of data that had been compiled by earlier researchers using the daily diary method. But we wanted to correlate these subjective responses with objective, time-locked physiological data, so on the first day we also outfitted each participant with a small cardiovascular monitor to wear at the waist, biosensors on the skin, and a blood-pressure cuff on the arm. Each time the beeper went off, and at the same time that they were to record their moods and activities, our study subjects pushed a button to activate the cardiovascular measurements.[8]

On the second day, instead of the heart monitoring equipment, each study subject carried salivettes, small rolls of gauze in sanitary containers. Each time the beeper went off, when they sat down to record their thoughts and experiences, they would chew the gauze to collect saliva, then return it to its container. This allowed us to correlate their levels of salivary cortisol, a marker for stress, with their self-reports of perceptions of loneliness and stressful or pleasurable activities. Over the next five days we continued to beep them nine times per day so that they could continue the questionnaires and journal entries.

Investigating the Inevitable

Many researchers trying to understand how the body experiences the wear and tear associated with aging limit themselves to cellular concerns. Some investigators study the molecular waste products called oxidants that gum up the cellular machinery. Others focus on the simple number of cell divisions taking place over time and the decline in transcription accuracy as each cell makes copy after copy of its DNA. Still others concentrate on telomeres, the protective coatings on the tips of chromosomes that gradually shorten with age.[9]

At the other end of the spectrum, researchers investigate the effects of social factors that can compound over time. The now-famous Whitehall study of British civil servants showed that people at each employment grade in government service experienced worse health and had higher mortality than those in the grade immediately above them.[10] This inequality persisted well into the middle-class range, so it was not just that higher-ups had better diet or better access to health care or better suspension systems in their cars. Somehow, the finely graded difference was created by social context—in this case, a rigidly hierarchical bureaucracy.

In the 1970s much attention was paid to the cardiovascular effects of being high strung. The hard-driving, competitive habits thought to cause heart attacks were labeled "Type A" behavior. Later research revealed that one factor within the Type A cluster—hostility—was the best variable for distinguishing those who would develop heart disease from those who would not.[11] This was a powerful finding for our own work, because, like loneliness, hostility is an attribute that can persist over time. Like loneliness, it is characterized by mistrust, cynicism, and feelings of anger that lead to antagonistic or aggressive behavior.[12] In one study of patients undergoing coronary angiography—a procedure in which an opaque dye is injected into the heart and then the heart's pumping action is recorded on film—merely recalling anger was sufficient to induce already diseased coronary arteries to constrict.[13] In patients with coronary heart disease, the recall of anger has been shown to produce an acute impairment of ventricular function as well.[14]

One hypothesis offered to account for these findings is that people who are hostile have exaggerated cardiovascular reactivity to stress, and that this heightened reactivity either contributes to the development of atherosclerosis or triggers a heart attack. But in a social context, hostility is also correlated with lower socioeconomic status, as it is with increased likelihood of smoking, with decreased likelihood of quitting smoking, and, as we saw in earlier chapters, with loneliness.

Did being on the lower rungs of the British civil service create hostility? Did it somehow create a sense of social isolation? Or was there some physiological effect that is common to the subservience and frustration that comes from working in a large organization and the subjective experience of loneliness?

Most behaviors are not randomly distributed but are socially patterned, meaning that they tend to occur in clusters. Many people who drink heavily also smoke. Those who eat a healthful diet also tend to exercise. Individuals influence and are influenced by their families, their social networks, the organizations in which they participate, their communities, and the society in which they live. This phenomenon came into sharp focus in the summer of 2007, when *The New England Journal of Medicine* published a study showing that—as the newspaper headlines would later simplify it—"our friends make us fat." By following the lives of 12,067 people over thirty-two years, researchers found that obesity occurred in social clusters.[15]

Social class also has its own distinctive role to play in health. People who have less money and less education endure more social stressors in the form of unemployment, work injury, and lack of control over their environment. They report fewer social supports, and they more often have a cynical or fatalistic outlook. Socioeconomic status is also strongly related to access to preventive care, ambulatory care, and high-technology procedures. People at the lower end of the socioeconomic scale are more likely to engage, not just in smoking, but in a wide array of other risky behaviors, and they are less likely to engage in health-promoting behaviors. In the language of epidemiology, this puts them at risk of having greater risks.[16]

But again, the Whitehall study did not find impaired health only at the bottom of the socioeconomic ladder. What sort of influence could be so subtle that it had a slightly greater effect at each step down the ladder of a multilevel hierarchy? The data showed that an imbalance in effort and reward and low levels of control in one's job were independent predictors of heart attack even when adjusted for age, employment grade, negative affectivity, and coronary risk profile.

The Need to Adjust

Part of the correlation between social stress and negative health outcomes seems to be the cost of maintaining physiological balance over time: our old friends regulation and co-regulation. As conditions fluctuate in the outside world—heat or cold, crisis or calm—

our bodies need to maintain a relatively stable environment, both within the organism as a whole and within each cell.[17] To keep us on a fairly even keel, we have control systems that adjust automatically. As the temperature outside fluctuates, for instance, our bodies adjust internal conditions to maintain a fairly constant temperature of approximately 98.6 degrees Fahrenheit. But there are limits. If external conditions are too extreme, the body's ability to adjust is overwhelmed, and we can die of heat stroke or hypothermia. There are also circumstances in which the body overrides standard operating procedure and allows temperature, or some other mechanism, to rise or fall a few degrees beyond the normal parameters. Most of the microorganisms that cause human disease cannot tolerate high temperatures, so part of our immune response is to elevate body temperature up to, say, 102 or 104—commonly called a fever—in an effort to ward off infectious agents.

The standard operating procedure—staying within the tightly prescribed boundaries—is called homeostasis. The process of making special adjustments of those boundaries according to the broader needs of the organism is called allostasis.[18] Homeostasis is like the first chair violinist: vitally important, but wedded to playing the musical score faithfully as it is written. Allostasis is more like a conductor, responsible for the entire orchestra, who brings a particular interpretation to the performance, and who may, within certain broader limits, alter the tempo or volume or other dynamics of the score as written by the composer.

Both ways of proceeding are necessary and desirable, but every time a body system responds to a stressor, whether the response is homeostatic or allostatic, there is a physiological cost to making the adjustment. The more complex the allostatic adjustment, the greater the number of bodily systems—endocrine, cardiovascular, immune—that become involved. The more systems, the larger the needed adjustments, and the more frequently these adjustments are required, the higher the physiological cost of bringing your body back to normal. The total cost of all these adjustments, the rigidly constrained and the more broadly orchestrated, is called the allostatic load.

Certain kinds of stressors can be beneficial, like the pruning that makes rose bushes or fruit trees more productive. The military makes boot camp training highly stressful in order to prepare the new recruit for battlefield conditions. When you go to the gym and lift weights, appropriately graduated levels of stress can make your muscles grow and increase bone density in the skeletal areas supporting those muscles.

However, the stress of having only limited power in a hierarchy, or the stress of feeling isolated in your community, your school, or your marriage, is not an experience that is going to enhance growth. Persistent social stress is not even likely to make you "strong at the broken places." What it does do is increase wear and tear throughout the system. After too many years of arranging all the parts in coordination with one another and with the changing outside conditions, the conductor leaves the podium exhausted.[19]

Looking at the question in a multivariate, multilevel way, we do not find a single, simple answer to the question of how loneliness causes ill health. Instead, the most accurate assessment is to say that it is a grinding process of wear and tear that proceeds along five intersecting pathways. It so happens that these five pathways sum up much of the physiological data we have explored so far.

Loneliness and Health: Five Causal Pathways

PATHWAY I: HEALTH BEHAVIORS

The social control hypothesis suggested that it was the absence of caring friends and family that led to people neglecting themselves or indulging themselves to the point of damaging their health. However, we found that the health-related behaviors of lonely young people were no worse than those of socially embedded young people. In terms of alcohol consumption, their behavior was, in fact, more restrained and healthful. Even among the older adults we studied, it was the subjective sense of loneliness—not a lack of objective social support—that *uniquely* predicted depressive symptoms, chronic health conditions, and elevated blood pressure.[20]

That being said, our study of older adults did indicate that, by middle age, time had taken its toll, and the health habits of the lonely had indeed become worse than those of socially embedded people of similar age and circumstances.[21] Although lonely young adults were no different from others in their exercise habits, measured either by frequency of activity or by total hours per week, the picture changed with our middle-aged and older population. Socially contented older adults were thirty-seven percent more likely than lonely older adults to have engaged in some type of vigorous physical activity in the previous two weeks. On average they exercised ten minutes more per day than their lonelier counterparts.

The same pattern held for diet. Among the young, eating habits did not differ substantially between the lonely and the nonlonely. However, among the older adults, loneliness was associated with the higher percentage of daily calories from fat that we noted earlier (and that is illustrated in Figure 6).[22]

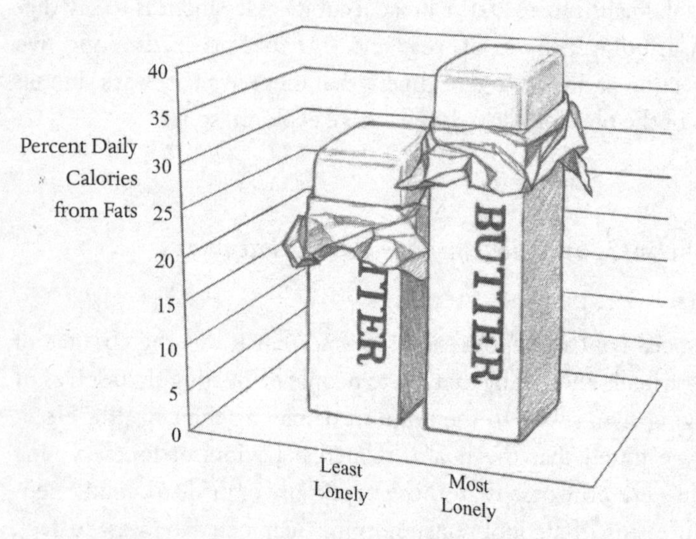

FIGURE 6. The U.S. Department of Health and Human Services and Department of Agriculture recommend that fat intake be between 20% and 35% of total calories. For participants in our study, an overall average of 34% of daily calories came from fats. However, the 20% of participants who were least lonely got only 29% of their calories from fats, while the 20% who were most lonely got 39% of their calories from fats.

It may be that the decline in healthful behavior in the lonely can be partially explained by the impairment in executive function, and therefore in self-regulation, that we saw in individuals induced to feel socially rejected. Doing what is good for you, rather than what merely feels good in the moment, requires disciplined self-regulation. Going for a run might feel good when you're finished, but for most of us, getting out the door in the first place requires an act of willpower. The executive control required for such discipline is compromised by loneliness, and loneliness also tends to lower self-esteem. If you perceive that others see you as worthless, you are more likely to engage in self-destructive behaviors and less likely to take good care of yourself.

Moreover, for lonely older adults, it appears that emotional distress about loneliness, combined with a decline in executive function, leads to attempts to manage mood by smoking, drinking, eating too much, or acting out sexually. Exercise would be a far better way to try to achieve a lift in mood, but disciplined exercise, again, requires executive control. Getting down to the gym or the yoga class three times a week also is much easier if you have friends you enjoy seeing there who reinforce your attempts to stay in shape.

So social environment definitely matters. It affects behavior by shaping norms, enforcing patterns of social control, providing or not providing opportunities to engage in particular behaviors, and either producing or reducing stress.

PATHWAY 2: EXPOSURE TO STRESSORS AND LIFE EVENTS

Our surveys with the undergraduates at Ohio State showed that lonely and nonlonely young adults did not differ in their exposure to major life stressors, or in the number of major changes they had endured in the previous twelve months. Our beeper study, in which we asked them to sit down and record their thoughts and experiences at various times during the day, also showed that there was no difference in the reported frequency of hassles or uplifts they experienced on a typical day, or in the number of minor irritants they were confronting when the beeper randomly interrupted them. At least for young adults, then, we found no evidence that loneliness increased exposure to objective causes of stress.[23]

However, among the older adults we studied, we found that those who were lonelier also reported larger numbers of objective stressors as being "current" in their lives. It appears that, over time, the "self-protective" behavior associated with loneliness leads to greater marital strife, more run-ins with neighbors, and more social problems overall.[24] Whereas socially contented adults reported, on average, 4.8 chronic stressors, lonelier adults reported 6.0, a twenty-five percent difference that then continues to compound over the course of a lifetime.

Similarly, this greater stress in the lives of the lonely may be compounded by a tendency to be trapped, like those in the middle rungs of the British civil service, in frustrating jobs. Perhaps because of their problematic social responses, people who get stuck in loneliness are less likely to get the top spot. The secondary slots to which the chronically lonely are often relegated can be psychologically and even intellectually demanding, but they allow the individual only limited control, a combination that has been implicated in high levels of job stress and adverse health outcomes.[25] What seems true for midlevel civil servants regarding a disconnect between effort and reward and a minimal control over one's circumstances also seems true for individuals trapped in a persistent feeling of social isolation.

PATHWAY 3: PERCEIVED STRESS AND COPING

Even setting aside the larger number of objective stressors in their lives, the lonely express greater feelings of helplessness and threat. In our studies, the lonely, both young and old, perceived the hassles and stresses of everyday life to be more severe than did their nonlonely counterparts, even though the objective stressors they encountered were essentially the same. Compounding the problem, the lonely found the small social uplifts of everyday life to be less intense and less gratifying.[26] The presence of and interaction with other people in their lives did not cause them to rate the severity of everyday stressors any less intensely. This finding dovetails with an fMRI study we will examine more closely in Chapter Nine that showed an anomaly in the way people who were lonely reacted to a picture of a happy human face. Ordinarily this sight activates a reward area of the brain, but loneliness dampens this response.

The extent to which we perceive experiences as stressful or as restorative has a profound influence on our health over time, but so does the way we respond. As I've mentioned, within reasonable limits—and despite all the self-help books written about its perils—a manageable level of stress can strengthen us, motivate us, and keep us on our toes. However, when people feel lonely, they are far less likely to see any given stressor as an invigorating challenge. Instead of responding with realistic optimism and active engagement, they tend to respond with pessimism and avoidance. They are more likely to cope passively, which means enduring without attempting to change the situation. This pattern of "grin and bear it" (while boiling inside) carries its own specific costs.[27]

Among young adults, we found that the greater the degree of loneliness, the more the individual withdrew from active engagement when faced with stressors. Similarly, the greater the loneliness, the less likely was the individual to seek either emotional support or instrumental (practical) support from others.[28] We found passive coping and refusal to seek emotional support common among lonely older adults as well.

PATHWAY 4: PHYSIOLOGICAL RESPONSE TO STRESS

The autonomic nervous system—the system that operates below the level of consciousness and governs physiological responses such as blood pressure regulation—(see Figure 7) is another of those "approach-withdraw" mechanisms in biology in which paired systems work in closely regulated opposition. Here, the gear for forward is the sympathetic nervous system. The gear for reverse is the parasympathetic nervous system. In response to a stressor, the sympathetic nervous system revs the engine—the heart, lungs, and other organs—priming them for immediate action: the fight-or-flight response. After that activation has served its purpose, the parasympathetic nervous system serves to relax those internal organs and allow them to back off.

In contemporary society, as we've noted, most of our stressors do not come and go in the kind of short, life-or-death confrontations that drove the evolution of "fight or flight." We can have the same overbearing boss, the same long commute, the same worries about health care and retirement, and the same feelings of social isolation, hour after hour, year after year. Moreover, we now experience these persistent stressors over a life span that, on average, extends well beyond what was the norm during all but the last few centuries of our species' existence. The environment is entirely different now than it was in our environment of evolutionary adaptation, but our

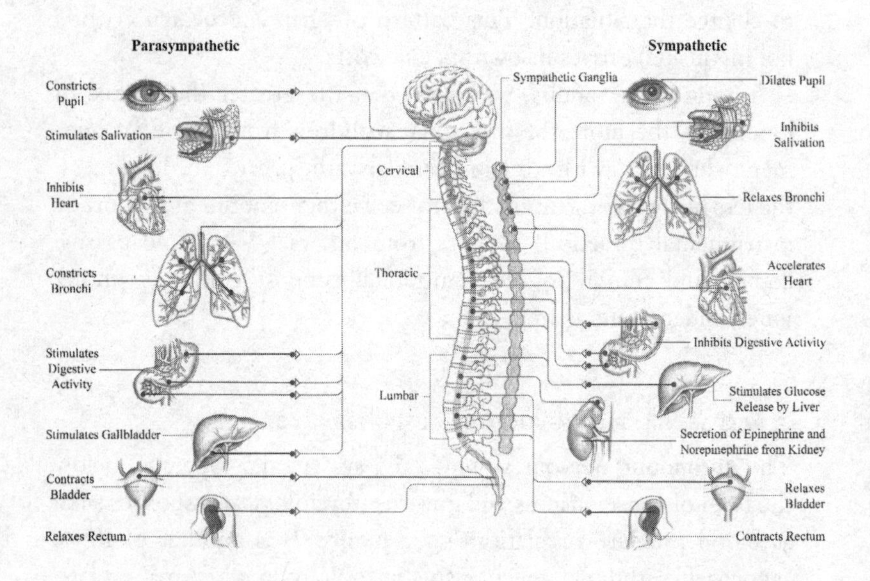

FIGURE 7. The autonomic nervous system directs all activities of the organs of the body that occur without a person's conscious control, such as breathing and digestion. It has two parts: the parasympathetic division *(depicted on the left),* which reduces physiological activation and helps conserve the body's energy, and the sympathetic division *(depicted on the right),* which increases physiological activation.

autonomic reactions have not changed. Our bodies still respond physiologically to the chronic, low-grade stressors of modern life as if we were fighting tooth and claw, and with no regard for conserving resources for the golf or tennis we hope to play when we're seventy or the nice long walks we want to take when we're eighty. Such an extreme response substantially exceeds the metabolic requirements of dealing with the stressors we're up against. All the same, every jolt into an unnecessarily high gear requires a compensatory down-shift, and all of these changes, up and down, year after year (the allostatic load), add up, exacting a high price for what amounts to very little benefit.[29]

We found loneliness to be associated with higher traces of the stress hormone epinephrine in the morning urine of older adults.[30] Other studies have shown that the allostatic load of feeling lonely also affects the body's immune and cardiovascular function. Years ago, a classic test with medical students showed that the stress of exams could have a dramatic dampening effect on the immune response, leaving the students more vulnerable to infections. Further studies showed that lonely students were far more adversely affected than those who felt socially contented.[31]

A natural part of the immune response is inflammation, the redness we associate with injury or infection. This often discomforting reaction actually helps bring in immune cells to fight bacterial invaders and to promote wound healing. But too long an exposure, or a dose of inflammation that comes too late, can slow the healing process, cause swelling and pain, and lead to loss of function in joints.[32] When inflammation is chronic it promotes cardiovascular disease.

Saliva samples allowed us to measure the morning rise in cortisol—the steroid that, in response to stress, acts on the body's metabolism and muscular efficiency to increase our ability to run fast, fight hard, or otherwise deal with physical threats. Cortisol makes us more alert, and it also adjusts our inflammatory and allergic responses to prepare the body to cope with potential injury.

Our studies showed that loneliness on a given day predicted a higher rise in cortisol the *next* morning. In addition, when we drew blood from our older adults and analyzed their white cells, we found that loneliness somehow penetrated the deepest recesses of the cell to alter the way genes were being expressed. Loneliness predicted changes in DNA transcription that, in turn, made changes in the cell's sensitivity to circulating cortisol, dampening the ability to shut off the inflammatory response.[33]

Loneliness may damage the cardiovascular system not just by inflicting stress, but also by promoting passive coping in the face of stress. Your blood exerts pressure within your circulatory system in much the same way that water exerts pressure in a garden hose. Increases in the "pounds per square inch" come either from pumping a greater volume of liquid into the same space or from "squeezing the hose"—reducing its interior diameter. The measurement of volume pumped by the heart each minute is called the cardiac output; the measurement of constriction in the small arteries—squeezing of the hose—is called the total peripheral resistance (TPR). Active coping—"taking arms against a sea of trouble"—raises blood pressure primarily by revving up cardiac output. Coping passively—which is what we do when we feel isolated—raises blood pressure primarily by constricting the small arteries, also known as increasing total peripheral resistance.[34] Higher TPR forces the heart muscle to work harder to distribute the same amount of liquid through the blood vessels. Meanwhile, the reduced diameter of and greater pressure in those blood vessels makes them more susceptible to wear and tear.

In our Ohio State studies we found that the greater the degree of loneliness of the student volunteers, the higher their TPR, even when their overall blood pressure remained normal. This was the case even when we added a bit of stress by asking the students to get up and speak in public, and also when we beeped them during the course of a normal day. In these situations, stress caused blood pressure, TPR, and cardiac output to go up for everyone. But because the lonely group started out with higher TPR, their TPR under stress was higher still.[35]

When we are young and resilient, like these undergrads, the

added wear and tear of higher TPR does not produce symptoms that a doctor would need to treat. But across the life span, for someone who remains lonely, there is a progression from innocuously higher TPR to high blood pressure that doctors would be concerned about.[36] At the same time, loneliness makes the lonely person less able to absorb the stress reducing (and TPR-lowering) benefits that others derive from the comfort and intimacy of their human contacts.

PATHWAY 5: REST AND RECUPERATION

Richard Lewontin suggested in *The Triple Helix* that we owe the gains in life expectancy over the past century not only to the public health measures we usually credit, things like better sanitation and better medical care. As much as anything, he says, we live longer because of improvements in living standards that allowed for greater recovery from the grinding stresses of life. "As people were better nourished and better clothed," he explains, "and had more rest time to recover from taxing labor, their bodies, being in a less stressed physiological state, were better able to recover from the further severe stress of infection."[37] It was also the five-day work week, then, and not just better medicines, that added years to the average life span. In the past few decades, though, our "can do" culture has reversed the trend toward increased leisure, placing a higher value on economic productivity than on recovery time spent with friends and family. Apparently the business and political leaders who drive our economy have not taken into consideration the economic impact, in terms of both lost productivity and health care costs, of this "rest when you die" attitude.

One clearly demonstrable consequence of social alienation and isolation for physiological resilience and recovery occurs in the context of the quintessential restorative behavior—sleep. Sleep deprivation, we know, has effects on metabolic, neural, and hormonal regulation that mimic those of aging.[38] At Ohio State, when we asked participants to wear a device called the "nightcap" to record changes in the depth and quality of their sleep, we found that total sleep time did not differ across the groups. However, lonely young adults reported taking longer to fall sleep and also feeling greater daytime fatigue.[39] Our studies of older adults yielded similar findings, and longitudinal analyses confirmed that it was loneliness specifically that was associated with changes in daytime fatigue. Even though the lonely got the same *quantity* of sleep as the nonlonely, their *quality* of sleep was greatly diminished.[40]

The Unit of One

Even in things medical, it seems, focusing on the individual without considering social context reveals only part of the story. As we have just seen, loneliness is a perpetrator that uses five different modus operandi to undermine our health. Once we see loneliness on the list of serious risk factors for illness and early death, right alongside smoking, obesity, and lack of exercise, that context should heighten our motivation to improve our level of social satisfaction, both as individuals and as a society.

But in a broader sense, the power we can now assign to loneliness, and conversely the power we can now assign to social connection, illuminates human nature itself, underscoring the "obligatorily gregarious" aspect of our species that is too often overlooked. Our sociality is central to who we are. Using loneliness as a window on human nature gets us beyond abstract (and therefore sometimes not terribly useful) debates about whether we are fundamentally Hobbesian beasts or "just a little lower than the angels." It also gives us a basis for ethics that transcends any particular religious or cultural tradition. In this framework, what is "good" pretty much aligns with what is good for our physiology, and also what is good for our species in the long run.

Just as the five pathways above are not activated magically, the most useful truths about who we are remain grounded in the demonstrable laws of nature. To fully appreciate the natural forces that created us, we need to dig a little deeper, to explore how and why social context—either loneliness or social contentment—has the ability to affect us so profoundly. The next few chapters will examine the neural and chemical mechanisms directly responsible, the ones that make trying to improve any individual's fate, including yours and mine, a cooperative venture.

PART TWO

from selfish genes to social beings

I do go for weeks isolating myself, not answering the phone, but then it seems I need to be touched. I now am more aware of it and sometimes I reach out and touch someone on the arm or hand, someone that seems to be hurting. I made a resolution last year to make more eye contact with people and say hello to strangers every day. I am surprised by their reaction. It is very uplifting for me and I hope for them.

—*Email from a woman in Florida*

CHAPTER SEVEN

sympathetic threads

"We cannot live only for ourselves," wrote the nineteenth-century preacher Henry Melvill (although the quote has been famously misattributed to Herman Melville, the author of *Moby-Dick*). "A thousand fibers connect us with our fellow men; and along these fibers, as sympathetic threads, our actions run as causes, and they come back to us as effects."[1]

When Melvill the preacher (not Melville the novelist) referred to threads and fibers, he was commenting on our power to influence others to stumble into sin or to live a righteous life. Spiritual guidance lies outside my expertise, but our scientific data tell me that the metaphor applies to interpersonal human behaviors, moment to moment, in myriad other ways.

In the last chapter, we saw that obesity occurs in social clusters.[2] If everybody we know seems to be putting on a few extra pounds, it makes it easier for us to accept our own added bulges. When everyone else looks a little heavier, our own fuller image in the mirror is less troubling, and it seems less likely that anyone else in this "expanding" circle is going to single us out for criticism.

But the limits of the "social control" hypothesis in explaining declines in health shows us that there is more to such social effects than purely social influences, something going on that is both deeper and more immediate than peer pressure or altered perceptions.

"Sympathetic threads" sounds vaguely mystical, and indeed similar ideas are deeply embedded in many religious traditions. Some people believe in the power of shamanism, others assign great power to voodoo. Chinese astrology specifies that a person's fate is influenced by the year of his or her birth. Westerners may dismiss all this as superstition, but there is no denying that these belief systems can have a tangible physiological effect—at least within the community of believers. One large study showed that Chinese Americans with a combination of a disease and a birth year that Chinese astrology regards as ill-fated were likely to die sooner than Caucasians who were similar in terms of age, health, and other lifestyle factors. The more strongly individuals were attached to Chinese traditions, the more years of life they lost.[3]

So the idea of sympathetic threads, or the idea that "our actions run as causes, and they come back to us as effects," cannot be entirely dismissed as hocus-pocus. There is such a thing as "cause at a distance" even within the highly rational realm of physics, and magnetism and gravity are two examples that come readily to mind. If you read about health in the newspapers, you are no doubt also familiar with the placebo effect, whereby patients respond positively to the actions of physicians even when those actions are neutral. Pills with no active ingredients—placebos—are administered as a control measure in clinical trials in order to measure the specific action of the new drug being tested versus the therapeutic benefit of simply interacting with the patient and appearing to do *something*—anything—for them. This is sometimes described as "mind over matter," sometimes dismissed with "it's all in your head." But in fact mind *is* matter, and there are very few mental activities in which interaction with the body is restricted to the cranial area.[4]

A couple of centuries after magnetism and gravity became firmly established in physics, the pioneering American psychologist William James (admittedly during a time of great public and even scientific curiosity about communication with the "spirit realm") began to investigate similarly unseen influences between and among living organisms.[5]

We don't have to do experiments in the lab (or conduct séances) to observe our physical responses being influenced by forces that we cannot see. Three times a day, most of us eat because it is "time to eat" as determined by the clock, the workday, or cultural tradition. We may have arranged to meet a client for lunch or a friend for dinner. And for most of us, somewhere between nine p.m. and midnight, it's time to turn out the lights and get some sleep, whether we feel drowsy or not. Even the most basic processes of eating and sleeping, then, are more than isolated chemical reactions—they are also responses to social conventions and social cues.[6]

Lab research, however, allows us to follow these social effects deeper into the organism. When the neurobiologist Suzanne Haber and the sociologist Patricia Barchas administered amphetamine to a group of male rhesus monkeys, they discovered that the drug had a widely varying—in fact, opposite—effect on individuals depending on their social rank. While the drug increased dominant behavior in males high in the social order, it increased *submissive* behavior in low-ranking males.[7] Social context determined the outcome of something otherwise thought to be a "purely" physiological reaction.

In trying to distance science from its murky medievalism, the seventeenth-century philosopher and mathematician René Descartes argued for a rigid distinction between rational processes and physical processes, between mind and body. But even Descartes saw a point of intersection. Erroneously, he theorized that animal spirits and "the rational soul" affected each other through the transfer of energy at the pineal gland, a pea-sized structure at the center of the brain. Assigned mystical powers in many traditions—Descartes saw it as "the seat of the soul"; in yoga it is the "sixth chakra" or the "third eye"—the pineal is actually an endocrine gland involved in the timing of many of our biological rhythms. Several decades ago, modern research showed the error in Descartes's distinction between mind and body (as well as his focus on the pineal). In just the past fifteen years, we have come to see that a rigid distinction between the world "out there" in the environment and the world "in here" within the mind/body is just as illusory.

In the 1980s, well removed from talk of animal spirits, neuroscientists introduced computer metaphors to talk about how the mind functions. But human intelligence is not something operating on the basis of closed circuits locked away inside the skull. If you want to create a brain as versatile as a human brain, its intelligence—like human intelligence—must be "embodied." This kind of information processing works from the ground up, through sensory (which means bodily) input. Rodney Brooks, who makes robots, gave up trying to make them smarter by focusing solely on symbolic processing—playing chess or doing advanced mathematics—the kind of tasks, as he told the *New York Times*, that "highly educated male scientists found challenging."[8] If you want to create a robot that can get along in the world, you need to give it the kind of capabilities that human toddlers need to master: knowing the difference between self and other, learning to interact with the physical environment, being able to distinguish between chalk and cheese.

When we analyze an object or a situation, we use both body and mind, integrating emotional, cognitive, behavioral, and neurophysiological processes. This "embodied" intelligence is also networked into perceptions and behaviors that synchronize, coordinate, and co-regulate with the perceptions and behaviors of other people. If our brains are like computers, then, they are not the "desktop box" of the 1980s, but the massively interconnected machines that share information and images on the World Wide Web. But even that metaphor does not do justice to the biological, which is to say the embodied, nature of our intelligence, or to the embodied nature of our social connections.

The Dancer from the Dance

The close integration of the development of mind and body, and of self and other, that is central to human experience begins within the womb. This tandem development then extends to a kind of imitative dance that continues to drive the anatomical and physiological development of our brains, "sympathetic threads" and all. This is an elaborate and absolutely essential choreography to which both the mother and the child contribute, and which goes on to shape to our response to social connection, as well as to feelings of isolation, later in life.

Newborns only hours after birth mimic certain facial behaviors. Open your mouth and they will open their mouth. Stick out your tongue and they will do same. Baby chimpanzees have the same ability, and both chimps and human babies stop doing this kind of facial mimicry at about the same age—two to three months.[9] Thus the window in time for this oh-so-tight connection closes just as human babies are ready to move up to the next stage of interaction—the ability to spontaneously vocalize and smile at other people.

Some infants—human, ape, or monkey—tend to imitate one kind of gesture, others a different kind. This variation aligns with innate temperament, which is linked to genetically programmed differences in sensorimotor systems, sensitivity, and reactivity that specifically govern varying degrees of responsiveness.[10] Some babies smile more than others; some are more easily startled. But imitation is an ability shared by all, one that promotes survival by "tuning" the infant to others, increasing attention from caretakers, and setting the stage for learning from these important people.

At six weeks, some babies are able to remember and imitate a gesture performed by an adult on the previous day—a skill that should help a baby identify the specific individuals vital to her survival. The importance of such interaction, and the close attention babies pay to parental faces, is indicated by the consternation they exhibit when confronted with an unresponsive caretaker. There is a vast literature detailing experiments in which parents present a "still face" to the baby and researchers monitor the baby's none-too-pleased reaction.[11]

The sympathetic threads of social connection and cause at a distance are so strong that we retain other forms of mimicry into adulthood. For instance, if you and I stand facing each other and I cross my arms, you are more likely to cross your arms. If you rub your nose, I am more likely to rub my nose. We adopt the speech patterns of others, and laughter or yawning can be contagious.[12] People even mimic the mannerisms of complete strangers, even when it is highly unlikely that there will be any future relationship or rapport.

Our physiology is tuned to others in ways we barely consider, but the depth and pervasiveness of the linkages suggest why frustration of the desire to connect can throw us into such a tailspin. At a sporting event, when we watch athletes with whom we identify, we lean toward them and unconsciously adopt their posture. The athletes themselves, if good enough at their game, are even more attuned to one another, anticipating teammates' moves as they work the ball downfield or execute the double play. Rapport contributes to synchronization, and synchronization contributes to rapport, which is one reason that a certain degree of compatibility, if not team spirit, is so important in sports, in a surgical suite, on the flight deck, or in the kitchen of a busy restaurant. Classes in which observers note a high degree of physical mimicry are the same classes which the students themselves rate as being high in rapport.[13] While people will mimic others to whom they are not particularly drawn, pairs of individuals who feel the most rapport are generally the most synchronized in their postures and movements.[14]

The old adage that imitation is the sincerest form of flattery seems to operate here, too. Participants whose postures had been mimicked—even when they had not consciously noticed the mimicry—later reported having a more favorable impression of the person doing the imitation.[15] Therapists know well that clients often rate their counselor more highly when the counselor has mimicked the client's postures. And identification or a desire to affiliate with an individual increases the degree of behavioral mimicry.[16]

In Chapter Three I described studies in which participants received a jolt to their sense of social well-being and, as a result, lost executive control. Other studies have induced the same feelings of rejection and then measured imitative behaviors. Observers noted that when recently rejected participants entered the presence of a seated person who was nervously shaking her foot, the participants dramatically—and unconsciously—increased their own foot shaking relative to their observed behavior in a similar situation prior to the rejection.[17]

Being rejected, especially being rejected by a group, lowers our self-esteem. It also makes us far more aware of social cues, forms of social information about group dynamics that might help us better navigate the social environment.[18] And yet, even as we become more attentive to facial expressions and vocal tone, feelings of rejection are associated with being less accurate in our interpretation of same. We apply more mental energy to the perception, but the added effort comes from a defensive, self-protective posture, which tends to distort that perception.

Rejected individuals also have a heightened tendency to conform to the opinions of others.[19] Do the self-proclaimed "ditto heads" who listen to Rush Limbaugh and other bombastic talk-show hosts suffer from feelings of social exclusion? It's a possibility. On the positive side, at least among women, those who had been ostracized or excluded in experimental situations were found to be more likely than others to contribute to a group task, even when their contribution would not be given individual recognition.[20]

Carrots and Sticks

Every living thing inherits systems of physiological carrots and sticks that direct its behavior. On a photo safari in Africa, a colleague of mine saw a pride of eight female lions moving across the plain in silent, subtle synchronization as they encircled a grazing herd of buffalo. He described for me the excitement of watching these big cats fan out across the open grassland. There was no obvious communication, but somehow each knew her role and carried it out in support of the larger mission, which was to isolate one of the weaker, slower buffalo, run it down, and kill it.

Our Paleolithic ancestors, even after they had developed speech, still needed this kind of wordless synchronization to bring down big game—or to trap rabbits, for that matter. Hunter-gatherer women, who contributed the bulk of the tribe's food while also tending to the children—the roots of multi-tasking—would also benefit to the extent that they shared an almost collective consciousness. This collective mind would include their cognitive understanding of the tasks at hand, their desire for social inclusion, and their sometimes equally subliminal fear of social exclusion. Do we know where all the children are? How far from camp should we roam? How widely can we afford to space ourselves without being endangered? How soon should we start back in order to reach camp before dark? Stressors in the environment accentuate this tendency to tend and befriend.[21]

The key concept for us is the extent to which the effortless sharing of knowledge or intuition relies on physical cues and sensations that are, themselves, imperceptible to our conscious minds.

For social behavior, the warmth of connection is the carrot; the pain of feeling isolated, also known as loneliness, is the stick. Our ability to have these sensations is embedded in the cells in our bodies, right down to the programming in our DNA, and yet at every level these physiological prompts are also mediated by the environment. Information about the environment conveyed by the senses, including information about our cultural and social environment, also affects the way our genetic blueprint will be expressed. Natural selection is the scorekeeper, determining which behaviors are adaptive and which are not, depending on differences in the rates of reproductive success among our surviving offspring.

Once again, while it exerts a powerful influence, DNA, sequestered as it is deep within the nucleus of each cell, has no direct contact with the world at large. DNA is in some ways like the mellifluous Charlie of *Charlie's Angels*, dependent on his minions to carry out his plans. For DNA the "minions" are not athletic young women played by film stars; they are the networks of biochemical and physiological functions which, collectively, we call the organism. These body systems, operating as intermediaries, facilitate the work of the genes that set the agenda, the individual's actions in the world, and the signals feeding back to the individual from the world. But just as Charlie's plans are a response to changing environmental demands, our genes are shaped by environments, past and present. This is why DNA is rarely the sole determinant of complex behavior, and why the genes that bias the intensity of our need for social connection do not completely determine our experience of loneliness.

In early infancy, a human newborn can be calmed by being placed against her mother's skin. In fact, it takes a while for an infant to discover that she and her mother are not actually one and the same. Reconciling this split between self and other—the desire for autonomy set against the desire for the calm assurance of intimate connection—remains a lifelong challenge, one that places a premium on self-regulation. And this regulatory balancing act between self and other, central to our experience of loneliness, is reflected in every cell in our body, because every cell reflects our evolutionary history.

The Lonely Cell

In different forms, this same dance back and forth across the boundaries of the organism—self and other, alone and connected—has been going on for as long as organisms have existed.

When life first emerged five billion years ago, what passed for "the organism" was not much more than a few molecules bound together into something we now call ribonucleic acid (RNA). What made this rudimentary life-form "life" (a status for RNA that not all scientists accept), was that it could store information and prompt biochemical reactions. Later a more sophisticated chemical compound, deoxyribonucleic acid (DNA), perfected a double-stranded informational code that it could rip down the middle, the two halves then reassembling in different combinations, which, because of this mix and match capability, would generate slightly modified copies of the original instructions, some of which might prove more adaptive than the original.

Over time, DNA began to provide instructions for assembling proteins into cell membranes, which created a boundary between self and non-self. From then on, the random process of evolution—

diversity, mixing and matching, competition, selection—favored organisms that could adjust conditions inside the self in response to changing conditions outside. This is the fundamental act of self-regulation that still persists within us, the quest for the balance. To achieve balance within the organism, individual cells have to coordinate. But individual organisms also coordinate with one another, as do aggregations of organisms, and on and on up the ladder to greater and greater levels of complexity, from beehives to book clubs. The molecular biologists Ned Wingreen and Simon Levin argue that the term "single-celled," even when applied to the amoeba, the classic creature studied by kids with microscopes, may be a misnomer. Even the lowly bacteria that cover our teeth form biofilms that are actually large interspecies collectives that provide benefit to us while taking care of their own. Similarly, four different species of bacteria live on the roots of tomato plants, working in coordinated fashion as they fix nitrogen, promote growth hormones, and fight off competitors. Again, there is no social contract—there is not even a coordinating intelligence—and yet these organisms have found a way to benefit from social connection and cooperation.

Sensing and Responding

More sophisticated social aggregations and their more sophisticated benefits required more sophisticated ways of using chemicals to sense and respond to one another. We see this next level of communication today when *Salmonella* bacteria rely on "quorum sensing" before getting down to business. This means that the bacteria remain dormant while they secrete small signaling molecules called autoinducers, which help determine the right moment to attack the host. That right moment is whenever these microscopic invaders reach sufficient population density to overwhelm the host's defenses. Similar systems for social coordination extend all the way to colonies of bioluminescent marine plankton that can act together like fans in a stadium who jump out of their seats and throw up their hands in a "wave." Acting in synchrony, the colonies luminesce when they sense the approach of a predator, lighting up the ocean surface as a form of collective defense. The light attracts larger predators, who, when the ploy succeeds, devour the would-be plankton-eaters.[22]

Along the path to more complex, multicellular organisms, such as you and me, communication and transport abilities continued to evolve. Eventually, the division between self and other ceased to be a question of two or more cells separated by a membrane. Self and other became two or more complex creatures separated by an exoskeleton, skin, feathers, fur, or scales. But the essential elements of social coordination, including sensing and responding, continued as before, involving physiological systems throughout the organism, from the level of cells, to individual organs, to organ systems, to whole creatures. Some of these now very complex organisms, in turn, organized themselves further into social collectives—hives, schools, flocks, and herds. The ability to regulate what went on inside the boundaries of each individual organism became even more complex, as did the ability of cells and systems within one organism to influence other cells and systems in another. Such social and physiological co-regulation allows bees to warm their hives, geese to fly south in protective formations, and certain fireflies seeking mates to flash in aggregations so large that they can be seen from space.

In any species in which social coordination advanced and persisted, it did so only because this kind of tightly regulated sensing and responding contributed to higher rates of reproductive success. Sometimes the concerted social behavior emerged from following a simple rule, such as "swim to the middle," which can serve the fish in the school the way "circle the wagons" served pioneers on the Oregon Trail. The fish in the middle are the ones who will survive to reproduce, and thus the genetically programmed behavior is passed along to their offspring.

In other social species, the key to collective action was chemical messengers called pheromones. Members of these species are equipped with chemoreceptor cells that can detect these trace elements either in the air or along so-called odor trails. The behavioral control provided by pheromones was able to create the intricacies of ant hills and termite mounds, but it could never have created the intricacies of London, Tokyo, or Mexico City. Chemical sensing requires that organisms stay in close proximity—so certainly it never could have led to international trade and vacations in Bali. But more to the point, the rigidity of a chemical system means that each of the co-regulating individuals is compelled to follow the instructions in exactly the same way. All the ants in an ant hill play by the rules all the time; not so all the humans in a big city, or a small village for that matter.

The Ability to Improvise

In becoming more sophisticated in their social interactions, higher organisms on the route to *Homo sapiens* transcended reliance on stock responses programmed in the genes or ingrained by way of parental behavior. Within certain general constraints imposed by the forces of natural selection, a far greater degree of individual improvisation emerged. Kittens and puppies do not blindly follow odor trails to the exclusion of other stimuli. Different breeds may have different characteristic behaviors, but individual animals are still free to scamper around and investigate all the stimuli in their immediate environment.

Vertebrates would continue to rely on pheromones as one of the sympathetic threads linking each to each, but natural selection gave the green light to a second means of communication that was hard-wired—a nervous system. Over time, this adaptation allowed for the reception of far more detailed information by the sensory organs, which led in time to responses that were in turn more subtle and complex. This new "sense and respond" system included a spinal cord and a clump of cells at the top that began to more closely coordinate the processes of communication and regulation. Early on, the range of action was still limited to the basic choice between approach and withdraw, and topics covered were still limited to the core business of biology—eating, mating, and trying to avoid pain.

Over the next several hundred million years, in the vertebrate brain, neural cells aggregated around the clump that sat atop the spinal cord. This became the brain stem of advancing species—sometimes called the reptilian brain because it first appeared in reptiles (see Figure 8). The outermost layer that eventually accrued to the more sophisticated brains of mammals is the cortex—sometimes called the neomammalian brain. Later still, a new or "neo" cortex evolved, which, in humans, has highly advanced prefrontal lobes that can formulate and interpret ever more complicated messages, symbolic messages such as $f(c) \# f(x)$ or "The unexamined life is not worth living."

But even among creatures with computational skills and a little knowledge of philosophy, primitive signaling through chemicals and other hard-wired behavior was never completely replaced; it was merely supplanted. This overlay of sophisticated over primitive further complicated our self-regulatory challenges—including those induced by loneliness. Having multiple operating systems made it not just possible but highly likely that, in the manner of Sheba the chimp facing down a dish of candies, we would be forced

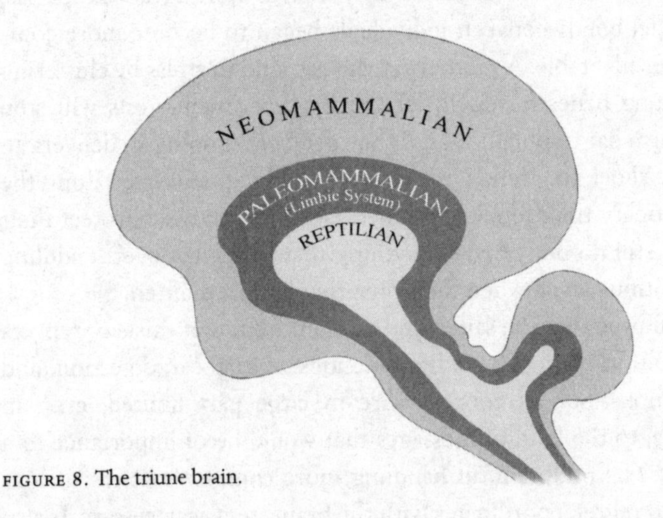

FIGURE 8. The triune brain.

to process simultaneous messages that were quite contradictory. "I want to eat that entire lemon meringue pie. I want to look good in my swimsuit when we go to Cancun."

Well before humans came along, even prior to the emergence of the cortex, a middle layer—the paleomammalian brain—evolved that added to the potential for confusion and complexity. Wrapped directly around the brain stem, and now with the later-evolving cortex wrapped around it, this "midbrain" is also called the limbic brain or sometimes the emotional brain. While it never gained command of numbers or words, it added to adaptability by increasing behavioral flexibility and contextual control. It is the peculiar partnering among these three layers—the animalistic brain stem that dates back to reptiles, the sometimes rational cortex, and the emotional brain caught in the middle—that allows lonely individuals to sometimes find themselves unaccountably screaming at their loved ones when what they really want is to be held.

Reptiles, which evolved before the emotional or midbrain came along, have a messaging system to link inner world and outer environment—including their interactions with other reptiles—that is not particularly fine-grained. Accordingly, reptiles are not known for their altruism, empathy, or parenting skills. With the advent of the paleomammalian (or emotional) brain in more advanced species, the social bonds between individuals began to become more complex and adaptable. A mother rat will respond to stress by clustering her young beneath her. In laboratory experiments, rats will stop pressing a bar to obtain food if they detect that doing so delivers an electric shock to a fellow rat nearby.[23] And yet, still later along the evolutionary trail, monkey mothers, who will actively protect their young, still do not provide anything that looks like overt cuddling and soothing, even when their offspring has been bitten.[24]

In all primates, the same kind of brain stem that exists in reptiles continues to control basic life functions—heartbeat, digestion, and respiration—but its concerns are in large part limited, even in humans, to the kind of messages that would be of importance to a reptile. The midbrain, in handling more complex matters such as love and regret, coordinates with the brain stem as necessary. It also functions, sometimes in coordination and sometimes at cross-purposes, with the most advanced part of the cortex, the frontal lobes that do the most sophisticated problem solving. And it is out of this complex system of sense and response, with multiple, sometimes conflicting levels of control, that the exhilarating sensations of social connection, and the devastating sensations of loneliness, began to emerge.

"Man still bears in his bodily frame the indelible stamp of his lowly origin," Charles Darwin wrote. To which the neurologist Antonio Damasio added, "The mind first had to be about the body or it could not have been."[25] Understanding the physicality of emotions is the only way we can fully appreciate how this form of cause at a distance—a subjective sense of well-being or distress based on our degree of connection to others—can exert the profound physiological effects that it does.

Looking more deeply at the invisible forces that link one human being to another helps us to see something even more profound: Our brains and bodies are designed to function in aggregates, not in isolation. That is the essence of an obligatorily gregarious species. The attempt to function in denial of our need for others, whether that need is great or small in any given individual, violates our design specifications. The effects on health are warning signs, similar to the "Check Engine" light that comes on in today's cars with their computerized sensors. But social connection is not just a lubricant that, like motor oil, prevents overheating and wear. Social connection is a fundamental part of the human operating (and organizing) system itself.

CHAPTER EIGHT

an indissociable organism

"All babies look like me," Winston Churchill once remarked, but it is evolutionary logic, not shared paternity, that accounts for their big round eyes and chubby cheeks. Natural selection gave babies the constellation of facial and vocal characteristics we classify as "cute" because "cuteness" promoted social connection. Cuteness was part of what made primeval mothers long to be with their babies. It also made fathers, grandparents—and today even passersby in grocery stores—want to interact with these miniature humans, amuse them, and protect them. There is now even a science of cuteness, as engineers in robotics try to make computerized companions that will have the same huggable appeal as a human baby.

Given that successful propagation of the genes requires survival of offspring, and given that human offspring not only are completely dependent but often awake and crying in the middle of the night, the bonding between human parents and children needs to be immediate and compelling. That is why, in this most fundamental form of human connection, there is far more going on than peek-a-boo, loving tolerance, the desire to protect, or preferences for a certain sound and a certain look. Researchers have used fMRI studies and essays to identify much of the specific brain circuitry and neurohormones associated with the motivation, attention, and empathy that are part of the parenting process.[1] But once again, the mind first has to be about the body, and much of the pleasure and pain that bond one person to another, in parent-child relationships, in sexual relationships, in all social relationships, are distributed throughout our anatomy.

In 1958, in a now-legendary, perhaps infamous experiment, the psychologist Harry Harlow of the University of Wisconsin removed newborn rhesus monkeys from their mothers. He presented these newborns instead with two surrogates, one made of wire and one made of cloth (see Figure 9). Either stand-in could be rigged with a milk bottle, but regardless of which "mother" provided food, infant monkeys spent most of their time clinging to the one made of cloth, running to it immediately when startled or upset. They visited the wire mother only when that surrogate provided food, and then, only for as long as it took to feed.[2]

Harlow found that monkeys deprived of tactile comfort showed significant delays in their progress, both mentally and emotionally. Those deprived of tactile comfort and also raised in isolation from other monkeys developed additional behavioral

FIGURE 9. In Harry Harlow's research, infant monkeys raised in isolation preferred the comfort of a cloth "mother" *(depicted on the left)* to a wire "mother" that provided food *(depicted on the right)*.

aberrations, often severe, from which they never recovered. Even after they had rejoined the troop, these deprived monkeys would sit alone and rock back and forth. They were overly aggressive with their playmates, and later in life they remained unable to form normal attachments. They were, in fact, socially inept—a deficiency that extended down into the most basic biological behaviors. If a socially deprived female was approached by a normal male during the time when hormones made her sexually receptive, she would squat on the floor rather than present her hindquarters. When a previously isolated male approached a receptive female, he would clasp her head instead of her hindquarters, then engage in pelvic thrusts.

Females raised in isolation became either incompetent or abusive mothers. Even monkeys raised in cages where they could see, smell, and hear—but not touch—other monkeys developed what the neuroscientist Mary Carlson has called an "autistic-like syndrome," with excessive grooming, self-clasping, social withdrawal, and rocking. As Carlson told a reporter, "You were not really a monkey unless you were raised in an interactive monkey environment."[3]

Harlow's experiment with monkeys would never be approved by a scientific review board today. Even in the 1950s, anyone attempting such brutality with human infants would have been arrested, and rightly so. Unfortunately, misguided social policies have shown us the effects that the "wire mother" of emotional deprivation and isolation—the imposition of loneliness at its most extreme—can have on children.

The most egregious example occurred in Romania, where the Communist dictator Nicolae Ceauşescu was a zealot for a coldly rational idea of technological progress. Before he was overthrown and shot, he had plans to raze every village in the country and replace all traditional housing with bleak Soviet-style apartment blocks. He did not live to carry out that scheme, but he did last long enough to execute a social vision that was even more pathological. In 1966 he outlawed contraception and abortion and instituted a system of rewards and medals to increase the birthrate. But he did nothing to help parents who were already economically strapped and unable to take care of the rising number of often unwanted children. Abandonment became rampant, with thousands of newborns consigned to orphanages that amounted to an emotional Gulag. As many as twenty children were in the charge of a single caretaker. There were no hugs, no laughter, no smiles, and certainly no lighthearted mimicry—the wide eyes and open mouths with which parents and babies charm each other and learn to bond.

When the orphanages were opened to the world in 1989, on the heels of Ceauşescu's downfall, outside health officials found three-year-olds who did not cry and did not speak. These children were in only the third to the tenth percentile for physical growth and grossly delayed in motor and mental development. Clutching themselves and silently rocking, they seemed to replicate the behavior of Harlow's socially deprived monkeys. Older children who had passed through this system in the late 1960s and early 1970s were still unable to form permanent attachments. Some found employment in the secret police, where their inability to care about others served the government's purposes. Others simply wandered the streets, homeless.

Once again, we see why human beings do not thrive as the "existential cowboys" that so much of modern thought celebrates. While it may be literally true that "we are born alone" and that "we die alone," connection not only helped make us who we are in evolutionary terms, it helps determine who we become as individuals. In both cases, human connections, mental health, physiological health, and emotional well-being are all inextricably linked.

Healthy Attachments

Healthy human development depends on the dynamics of early attachment, the bond between the baby and her caregivers. But attachments that are durable and healthy also depend on the innate psychological attributes of the infant, as well as the attributes of the adults (and sometimes older siblings) providing care. The quality of the caretaking bond provides the first instance in which genetic bias meets environment, and in which a child's subjective need for a feeling of connection will or will not be met. Early interaction with caregivers not only helps shape the infant's brain, but determines in large part how she will react to stress, including social stress. This is the moment when the first jolt of loneliness will or will not be introduced.

In trying to explain the dynamics of bonding in individual human development, the psychoanalyst John Bowlby was inspired a half-century ago by the work of Konrad Lorenz. This legendary student of ethology, or animal behavior, had shown that baby geese, which normally "imprint" the image of their mother and instinctively follow her, could just as easily imprint the image of, and line up behind, Konrad Lorenz.[4] Photographs of goslings following the Austrian scientist appeared all over the world. Building in part on Lorenz's findings, Bowlby developed what he called "attachment theory," which held that innate signals between parent and child shaped not only their relationship but the child's later personality. The primary shortcoming in Bowlby's theory is that humans are not geese.

A student of Bowlby's, Mary Ainsworth, subjected babies to a test she called the "strange situation," a series of eight short episodes of separation from and reunion with their mothers. On the basis of her observations of the infants' reactions, she developed three categories of what she called "attachment style." Children in the first category were blasé about their mother's departure and happy upon reunion. Those in the second group were not distressed when their mother left but shunned her when she returned. Those in the third group were terribly anxious when left alone and angry and upset upon reunion.[5] Ainsworth labeled the first group securely attached, the second group insecurely attached, and the third group anxiously attached. While attachment theory was highly influential in academic circles and popularized in parenting books by the pediatrician William Sears, researchers have criticized Ainsworth's studies for numerous shortcomings, including a small sample size and the problem of subjective evaluation.

More recent research has shown that levels and types of attachment in adults do not rigidly follow these experiences from childhood, nor do they necessarily remain constant from one relationship to another. It's not that childhood attachment doesn't matter; it's just that many, many other factors also matter—including the genetic propensity that sets the thermostat for feelings of loneliness, making the individual crave social connection a little or a lot.

Bowlby's conception of attachment is also limited by his focus on the individual in isolation, but since relationships inevitably consist of more than one person, each relationship creates its own dynamic. As our many examples of self-regulation, co-regulation, and even mimicry have demonstrated, each person involved in a relationship unconsciously influences the other.

The Role of Temperament

The psychologist Jerome Kagan largely dismissed attachment theory and took another tack. He saw the child's temperament, heavily biased by the genes, as the lead actor. In 1986 Kagan and his colleagues began a longitudinal study in which they dangled novel toys and otherwise presented unfamiliar stimuli to a group of five hundred infants. Twenty percent of these babies cried and vigorously protested, and Kagan labeled them "high reactive." Forty percent showed very little response at all, and thus they were "low reactive." Another forty percent were somewhere in the middle.

Kagan was able to bring many of these children back into his lab at regular intervals for follow-up studies. Somewhere between ages ten and twelve, almost half of the original group were given a full battery of brain scans and other clinical measures. In an effort to replicate the early experiment with the toys, some of these kids were asked to give an impromptu speech—a source of stress for most people. Twenty percent of those who had been labeled "high reactive" as infants still showed considerable distress in response to the performance pressure. One-third of the "low reactives" showed the same remarkable calm they had shown in early childhood. The vast majority had drifted into the middle range, while only five percent had switched from high to low or from low to high.[6]

The stability of reactivity over time underscores the importance of the genes in determining human personality, but it also reminds us of E. O. Wilson's description of DNA as an elastic leash. Kagan's results indicate a strong genetic influence, but the variability that emerged also underscored the role of the environment interacting with those genes.

The prevailing view today—the ethological theory—actually harks back to the work of researchers like Lorenz who studied animal behavior in its evolutionary context. Within the theoretical context of ethology and evolutionary psychology, developmental psychologists now study human behaviors in terms of their adaptive value, that is, their contribution to the propagation of genes. Our view of loneliness as an adaptive behavior that prompts social connection is, in this sense, ethological.

In an evolutionary context we can see that secure attachment, and the willingness to venture forth that it supports, is a good strategy for environments in which parents have the time and resources to attend to children. A safe and secure environment—emotional security as well as physical security—means that a child can explore without great risk. In a more stressed environment, whether it is the Kalahari Desert or the slums of Paris, the demands of survival can overwhelm parental attention, a situation that favors insecure attachments that keep children close. But a child's temperament can also alter the parental side of the equation.

Some children are good natured and easy to be around; others, through no fault of their own, seem to emerge from the womb cranky, demanding, and difficult. Some seem to giggle incessantly, others cry more than, understandably, their parents think fair. Some crave cuddling and attention at every moment; others, as infants, are happy to lie in their cribs playing with their toes, and then as toddlers explore quietly on their own. Some individuals are, as a rule, simply less positive and cheerful than others, a genetic bias that appears to be related to which side of the individual's brain is generally more active. Negative emotions such as fear and disgust activate the right prefrontal regions more than the left.[7] Stimuli that evoke positive emotions activate the left prefrontal regions more than the right.[8] Individuals with greater baseline activation on the right side are more likely to be withdrawn,[9] or to display greater negative affect as their everyday, run-of-the-mill emotional state.[10] Children who display social competence show greater left frontal activation.

Just as different children present different challenges, some parents are temperamentally attentive to their child's every smile or frown, while others range from blasé, to mildly disengaged, to absent, to abusive. Some parents are blessed with a well-ordered existence and a cheerful outlook, while others struggle to cope with their own needs, thus having far fewer emotional resources available for an "easy" baby, much less a fussy one.

The full range of human characteristics contributes to a complex dynamic that generates the adult's response to the child, which completes a feedback loop that will help shape the child's self-image. That dynamic also plays a role in the first experience of loneliness. A healthy match between parent and child can lead to a feeling that the world is your oyster, that you fundamentally belong, and that the essence of who you are is really A-okay. A mismatch can make you feel marginal. In the classic film *Five Easy Pieces*, the young man played by Jack Nicholson has a drunken encounter with a sad young woman played by Sally Struthers. She points to the dimple in her chin, then says her mother told her that it came from God, who when he first saw her coming down the assembly line, pushed her away as a reject. After such an abusive comment from a parent, it is little wonder that someone might grow up feeling somehow flawed and fundamentally alone.

Minding the Body

No matter which theory we embrace for how it came to be, our individual social orientation—including the intensity of our individual need for connection and our pain and disruption when that need is not met—is obviously intertwined with our physiology at the deepest levels.

In his introduction to *Descartes' Error*, the neuroscientist Antonio

Damasio offered three fundamental concepts that drive home this fact:

> (1) The human brain and the rest of the body constitute an indissociable organism, integrated by means of mutually interactive biochemical and neural regulatory circuits (including endocrine, immune, and autonomic neural components); (2) The organism interacts with the environment as an ensemble: the interaction is neither of the body alone nor of the brain alone; (3) The physiological operations that we call mind are derived from the structural and functional ensemble rather than from the brain alone: mental phenomena can be fully understood only in the context of an organism's interacting in an environment. That the environment is, in part, a product of the organism's activity itself merely underscores the complexity of interactions we must take into account.[11]

The psychologist Martha McClintock was still an undergraduate when she uncovered evidence of social influences at work on the endocrine systems of the young women living together in her college dormitory. As the semester wore on, she noticed that her dorm mates began to have their menstrual cycles more and more on the same schedule.

McClintock went on to show, in work with lab rats, that social animals that were housed together responded to a "ceremonial" signal marking the optimal time for conception. By confining the animals separately but having them share an air supply, she demonstrated that the source for the co-regulation is pheromones, the same airborne chemical signals, carried in concentrations below our threshold for smell, that direct many primitive behaviors. Rats born at the socially opportune moment during a synchronized wave of reproduction survive at a rate of eighty to ninety percent; those born out of step with the group survive at a rate of only thirty percent. Certainly in this instance, referring to the power of social connection as a matter of life or death is no exaggeration.

Similar social-biochemical effects determine, not who lives or dies, but who reproduces—which has an even more direct effect on subsequent evolution. Among Norway rats, males ejaculate more sperm when copulating in the presence of male rivals, seemingly because the competition to reproduce persists all the way up the fallopian tube to the surface of the egg. For the same reason, an ape's testicles are proportionate to the size of the male breeding pool. The male chimpanzee, surrounded by ruthless competition, has reproductive equipment that is truly prodigious, while the gorilla, living with a harem as the only male, has nothing to brag about. The evolutionary reason: A male without rivals needs no special adaptations to increase his odds of becoming a father.

Once again, the social and the physiological cannot be separated any more than we can separate the length from the width of a rectangle.

The Chemistry of Connection

In 1906 a young physiologist named Henry Dale, later Sir Henry Dale, first isolated a small protein, or peptide, associated with breastfeeding in all mammals. Arising from the pituitary gland, this substance, called oxytocin, also played a role in pacing the birth process. Later research would demonstrate that oxytocin is, in a way, the "master chemical" of social connection, and as close as anything we know to the love potions popular in romantic folk tales. Physical manifestations of connection such as hugs and back rubs increase oxytocin levels in the areas being touched.

When an infant suckles at the breast, the stimulation increases the concentration of oxytocin, which then hastens the release of milk. Over time this stimulation becomes a conditioned reflex for the mother, and simply seeing her baby is enough to cause her milk to let down.

Ewes that are given an injection of oxytocin will develop a maternal bond with lambs that are not their own. Ewes that are given a substance that blocks oxytocin during the birth process do not bond even with their own natural offspring. In the prairie vole, a female injected with oxytocin in the presence of a certain male will, from that moment on, recognize that male and prefer him over all others. These voles are highly social small mammals that form long-lasting bonds between mates. In contrast, their close relatives meadow and montane voles are mostly solitary creatures that do not form enduring bonds. Part of the difference in their relative need for social connection is oxytocin. The highly social and monogamous prairie voles have receptors for the substance concentrated in a specific reward area of the brain. The solitary montane and meadow voles do not.[12]

Like most chemicals in the body, oxytocin has an "antagonist" compound that works with it in coordinated opposition. Like rocket thrusters going on and off to steer a spacecraft, these two compounds steer behavior with infusions of chemical stimulation. Oxytocin's partner is vasopressin, a hormone that contributes to social bonding and in males stimulates aggression toward other males. Vasopressin makes female laboratory rats afraid of strangers, including young rats not their own. Oxytocin overrides even ordinary inhibitions against approaching another animal. After an injection of oxytocin, female lab rats begin to exhibit maternal behavior even when they have not been pregnant. They build a nest, pick up any young they find in the vicinity, carry them to the nest, then lick and groom them. Even though these artificially stimulated rats produce no milk, they still lie down as if preparing to nurse. They also defend their "adopted" children against other rats.

Oxytocin aids social regulation by being the chemical of calm. Apes spend ten percent of their waking hours picking through each other's fur, but this behavior is not just about hygiene, or even social courtesy and deference. The extended rhythmic touching involved in apes' grooming behavior stimulates the release of oxytocin, which helps promote social harmony.[13] Skeptics may be right to deride the idea that "all the world needs" is a big warm hug. Then again, who is to say that more hugging and less hitting could not have a positive effect in reducing antisocial behavior of all types?

The chemistry of social regulation, however, need not involve touch. Sympathetic threads in chemical form make a chimp's hair stand on end when he sees his rival and a young woman's heart race when she sees her beloved. These forms of cause at a distance begin with visual arousal, followed by the brain's instructions to release the stimulant norepinephrine.

Healing Calm

In Chapter Six I described the pathway of "rest and digest" that allows social connection to enhance health. In Chapter Seven I used the image of an automobile's master computer to convey the powerfully integrative force of social connection in our lives. Oxytocin operates one step closer to where the rubber meets the road, functioning like a car's timing belt, specifically linking many different systems under the hood. At this functional level it serves both as a hormone, communicating among organ systems, and as a neurotransmitter, signaling within the brain and throughout the autonomic nervous system.

In other ways as well, this "master chemical" of social connection and coordination acts with a uniquely high level of coordination. In oxytocin-producing cells, the electrical impulses that activate function do not come one by one, but occur in a cluster. When someone with whom we have a warm and personal social connection gives us a back rub, the coordinated electrical activity then causes the oxytocin-producing cells to act in concert, after which one good thing can lead to another. Positive sensations in the neck and back have a feedback effect that leads to increased oxytocin production, which promotes further social bonding, which can create opportunities for further, far more dramatic infusions of pleasure, and thus even more oxytocin.[14]

Areas of the brain influenced by oxytocin include the amygdala, the hypothalamus, and regions in the brain stem associated with the regulation of blood pressure, pulse, alertness, movement, and feeling. The same nerves connect with locations in the brain and spinal cord that control both the autonomic nervous system and the sensation of pain. This is why any encounter that stimulates the release of oxytocin can improve so many different aspects of our outlook as well as our physical well-being.

Hormones generally circulate throughout the entire body via the bloodstream. Individual nerves, which likewise inform and coordinate, reach only a limited area, where they deliver signaling substances that cause specific localized effects. As a hormone, oxytocin is synthesized in neurons of the hypothalamus, then released into the bloodstream through the pituitary gland. As a neurotransmitter, it can be delivered directly from the hypothalamus to the nervous system through long nerve fibers (see Figure 10).

The closeness inspired by oxytocin releases more oxytocin, which can serve to further reinforce group cohesion. Working up a sweat through exercise can promote the release of this bonding chemical, which may be one of the reasons that elated members of the win-

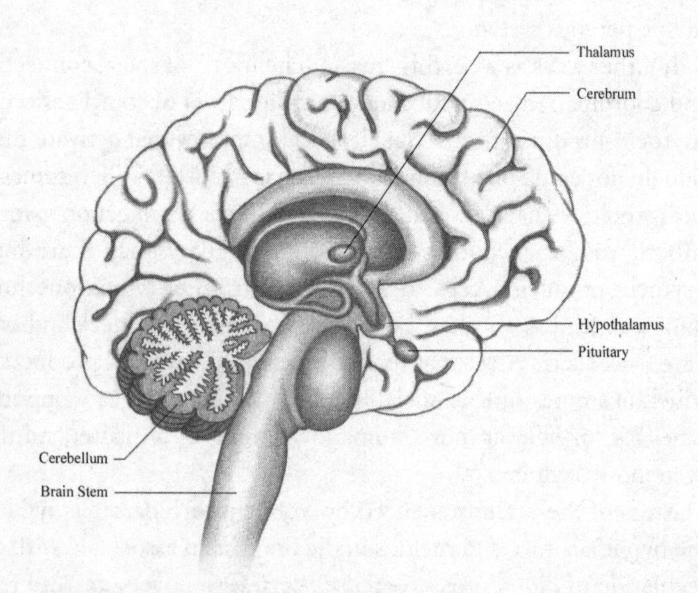

FIGURE 10. The hypothalamus and the pituitary gland are located next to each other deep within the middle of the brain.

ning chess team rarely hug each other and slap each other on the butt quite so enthusiastically as elated members of the winning hockey team.

Whereas loneliness inflicts pain, increases perception of stress, interferes with immune function, and impairs cognitive function, oxytocin (released when your spouse takes your hand as the plane charges down the runway) can reduce stress reactivity, increase tolerance for pain (as when Mommy 'kisses it to make it well'), and reduce distractibility (as when the coach grips your shoulder as she gives you instructions). In the laboratory, oxytocin enables experimental animals to keep right on with their maternal activities even in the presence of noise and bright light.[15]

We saw earlier that loneliness leads to a decline in executive control. Studies of children in daycare show that those who receive regular massages are calmer and better behaved. Perhaps because massage releases oxytocin in the masseuse as well, massage therapists, as an occupational group, typically show relatively low levels of stress hormones. They also tend to have blood pressure in the healthy low range. And the good feeling inspired by touch causes people to think of their masseuse as being trustworthy.[16]

When humans engage in the ultimate social connection of making love, orgasm releases a flood of oxytocin into the bloodstream, inducing calm—even drowsiness—as well as the same concentrated focus we see in nursing mothers. Here, too, the chemical of connection lowers blood pressure and the levels of stress hormones. Repeated over time, this experience of closeness helps create and maintain the pair bond so essential to our advance as a species, which is more than the rosy feeling we call "love." It is also an unconscious, physiological link that, temporarily at least, can mask the other ways in which a couple may be entirely incompatible. ("Opposites attract," an old adage says, "and then they attack.") This unconscious bonding is one of the underappreciated reasons why having sex with someone you're not sure about can be a bad idea. As in the case of the prairie vole, the chemical infusion can create a fixation on a single individual that otherwise may not make much sense.

Human Valences

Whether it is based on physical chemistry or on a shared passion for Hong Kong action pictures, the rapturous obsession we associate with young lovers many have speculated, often has an expiration date some three to seven years from the time of onset, more or less the time it takes for a couple's first child to move beyond the period of its most complete dependency.[17] Happily for the institution of marriage, three to seven years is long enough to establish other forms of affection, trust, and family bonding that can last a lifetime. More happily still, most people assess compatibility on the basis of deeper psychological dimensions, which can make simpler and more certain the task of achieving a long and satisfying marriage.

We speak of loneliness sometimes as being "out in the cold," and of the feeling we get from satisfying social connections as "warmth." Oxytocin creates literal warmth between creatures, in part, by redirecting warmth from one body region to another. Breastfeeding infants show increased blood flow in their hands and feet. The warmer the mother, the warmer the baby's feet. Oxytocin creates the same redirection of temperature in any number of human encounters in which we see warm chests and rosy cheeks, whether it is mothers nursing, fathers holding their babies, or lovers entwined in a postcoital nap.

The lining of the digestive system has the same developmental roots as the skin, so it is not entirely surprising that eating serves as a kind of internal massage that also stimulates the release of oxytocin. All food, then, is to some extent comfort food, and a good meal with good friends is the best of both worlds when it comes to relieving stress. We may overeat when lonely, in part, because the feeling of isolation has impaired executive control, but it is also true that eating simply feels good. Eating is a way of self-soothing that carries costs when we take it to excess, but that does not make it any less soothing in the moment. When we are pleasantly full, our problems seem more distant, and we feel closer to the people around us. In a similar vein, moderate amounts of alcohol increase the concentration of oxytocin in the blood, contributing to the conviviality of social drinkers. High amounts of alcohol have the opposite effect, which may contribute to the belligerence and antisocial behavior of obnoxious drunks.

Just as a mother's oxytocin level is increased by her infant's suckling, the oral gratification of sucking itself increases the baby's oxytocin levels and feelings of attachment. This comforting effect is why children can self-soothe by sucking their thumbs, or Daddy's pinky finger, or a pacifier. In adults, the release of oxytocin triggered by sucking may contribute to the addictive quality of smoking, as well as to the immediate intimacy often shown by smokers who may have nothing else in common, yet who will almost always give a cigarette or a light to anyone who asks. A cigar-smoking friend of mine said that part of the appeal of this otherwise nasty habit was that he never felt lonely when puffing on his stogie.

Part of the unfairness of loneliness is that it often deprives us of touch and the soothing comfort that it brings. But as we have seen, unwanted isolation in any of its forms—physical, emotional, spiritual—is deeply disruptive to an organism designed by nature to function in a social setting.

Earlier I described human beings as obligatorily gregarious creatures and mentioned this as the number-one consideration facing a hypothetical zookeeper trying to develop a proper enclosure for a member of our species. Unfortunately, about one hundred years ago, that thought experiment was made a reality.

In 1904 a young man named Oto Benga, a member of the Batwa People of the Congo, often called pygmies, was brought to the United States by a missionary and exhibited at the World's Fair. While Benga was away, word came back from Africa that his tribe had been wiped out, so when the fair closed, he was assisted by an orphan's home, then later moved to New York, where he was put on exhibit at the monkey house at the Bronx Zoo. Caged with a chimpanzee, he practiced with his bow and arrow, slept in a hammock, and was viewed by as many as forty thousand people a day. Protests by African-American clergy freed him from the cage, whereupon he was free to roam the zoo property as a kind of interactive exhibit. By all accounts already suffering from cultural dislocation, he then had dure the taunts of gawking zoogoers. Not surprisingly, he became slightly erratic in his behavior. Eventually, he was sent to another orphanage, and then down to Lynchburg, Virginia, where he went to work in a tobacco factory. Church groups tried to educate him; they even had a dentist cap his teeth, which had been filed to sharp points in the tradition of his people. But his alienation and erratic behavior persisted. According to observers' reports, he despaired that he would never be able to return to his home. In 1916, after twelve years of forced exile and humiliation, Oto Benga borrowed a revolver, pulled the caps off his teeth, built a ceremonial fire, and shot himself through the heart.[18]

"People must belong to a tribe," E. O. Wilson tells us; "they yearn to have a purpose larger than themselves." Social isolation deprives us of both our feeling of tribal connection and our sense of purpose. On both counts, the results can be devastating, not only for the individual, but for societies as well.

CHAPTER NINE

knowing thyself, among others

Almost two hundred years ago, Charles Darwin made a life-changing visit to the Galapagos Islands. On these isolated rocks off the northwest coast of South America he encountered varieties of life he had never imagined. His observations inspired him to think long and hard about diversity, competition, and change, ruminations that allowed him to develop the theory of evolution through natural selection. But Darwin was able to visit such an exotic locale only because he was a guest on the naval survey ship HMS *Beagle*. And Darwin was on that ship only because its captain, Robert Fitzroy, desired companionship, and was precluded by his rank from socializing with the other members of the ship's crew. In other words, Darwin might never have arrived at the major structural principle in our modern scientific understanding of life had it not been for the very human problem of loneliness.

In 1839 Darwin published an account of his adventures that became known as *The Voyage of the Beagle*. In 1859, after years of anguish about the religious and cultural implications of his ideas (and then only because a competitor, Alfred Russel Wallace, was nipping at his heels), he published his primary account of natural selection, *The Origin of Species*. Little more than a decade later, in 1872, he turned to the issues of human psychology in his last major work, *The Expression of the Emotions in Man and Animals*. But it was in his notebooks rather than in his published writings that he expressed the deeper question that remains central to our effort to understand the particular "cause at a distance" that underlies our most intimate and powerful social connections.

How is it, Darwin wondered, that a man's "kindness to wife and children would give him pleasure, without any regard to his own interest?" Pleasure is physiological. It suggests the satisfaction of physical appetites, such as the taste of a good steak or the warm feeling of sunshine on our skin. Kindness to others, like other social behaviors, can appear abstract, far removed from cells, body systems, and nutrients. But as we've seen again and again, the pain of loneliness and the uplift of connection are both deeply physiological. Both engage "sympathetic threads" in the form of the sensory responses and chemical reactions we've just explored.

As to the second half of Darwin's question, how is it that a parent who has to work three jobs to put a child through college can find sufficient pleasure in the exchange to say "I'd do it all again"? Then again, looking at the pleasure/self-interest equation from the other direction, how is it that being distanced from loved ones can give us emotional pain, even when the distance *is* serving our own interest, whether that interest is traveling on business or leaving the kids with a sitter while Mom and Dad have an all too rare evening out? Moreover, how do we decide what to do in the face of these competing desires? Satisfaction of the appetites versus satisfaction of meaningful goals? The pleasure of professional accomplishment versus the pleasure of friends and family; the pleasure of dinner and a movie versus the pleasure of putting the kids to bed?

The fact that our emotions are physical sensations does not mean that we are entirely at their mercy. European turnsoles are plants that respond to sunlight in a way that is invariable. They are heliotropes, meaning that they open and move toward the sun's rays each and every time those rays reach the plant's surface. That genetically programmed, invariable, reflexive movement toward a stimulus is called tropism.

Humans have certain reflexive actions—the doctor whacking you just below the knee with her little rubber hammer is a test of one such reaction—but we are not tropistic. We are able to exercise considerable discretion about how and when and why we do what we do, and this discretion operates at the intersection of emotional sensations, rational thought, and social behavior. If we had an invariable reflex to do "the right thing," it's doubtful we would assign much virtue to it. (It's also doubtful that we would have any need for a philosophical concept called "virtue.")

We humans take great pride in our intelligence—our capacity for rational thought—as what sets us apart from the rest of nature, and yet we usually gauge an individual's virtue or "humanity" not in terms of brain power but in terms of his or her emotional sensitivity. Most of us devote a tremendous amount of energy to trying to understand one another and trying to "do right" by one another, and even scam artists, swindlers, and self-serving politicians know they have to at least give lip service to the ideals of empathy and compassion.

Displaying the appropriate emotional response allows us to be accepted as properly human, but even here we humans are not unique. Even here, the roots go deep into the biology of our cells, and deep into evolutionary history.

Darwin approached the problem of emotional responses to pleasure and pain through the lens of social signaling. He hypothesized that being able to, in some sense, share their inner states with one other would make animals better able to anticipate, prepare, and perhaps coordinate their activities. Once signaling was in place, and despite its considerable contribution to survival, this expressive ability expanded beyond its functional benefits and began to be applied in ways that extended beyond the immediate business of staying alive.

Social Signals

In *The Expression of the Emotions in Man and Animals*, Darwin related the story of two chimpanzees as told to him by a zookeeper: "They sat opposite, touching each other with their much protruded lips; and the one put his hand on the shoulder of the other. They then mutually folded each other in their arms. Afterwards they stood up, each with one arm on the shoulder of the other, lifted up their heads, opened their mouths, and yelled with delight."[1]

Frans de Waal of Emory University, director of the Living Links Center at the Yerkes Primate Center in Atlanta, Georgia, is perhaps the leading proponent of acknowledging what we might call the "inner lives" of animals. He tells a story of two chimpanzees shut out of their zoo shelter during a rainstorm, then discovered by the primatologist Wolfgang Kohler. Seeing the drenched and shivering animals, Kohler opened the door for them, but instead of scurrying past to get inside, both stopped to give him enthusiastic hugs.[2]

Do animals experience joy in companionship the same way we do? Do they need contact with their mates or buddies the same way we do, and thus can they, too, sometimes feel woefully isolated?

In his book *Our Inner Ape*, de Waal describes a female gorilla named Binti Jua, who became a world celebrity when she rescued a young boy who had fallen into the primate exhibit at Chicago's Brookfield Zoo. De Waal also tells of a bonobo—a species closely related both to chimps and to us—who saw a starling hit the glass of her enclosure at a zoo in England. The bird was stunned, and Kuni, the bonobo, gently set it on its feet. When it did not recover, she threw it, its wings fluttering ever so slightly. Then Kuni climbed to the top of the tallest tree in her enclosure, unfolded the bird's wings, and tossed it like a paper airplane. Nothing seemed to work, but Kuni kept watch over the bird. Her efforts, however vigorous, evidently did not hurt the starling; by the end of the day it had recovered and flown away.

De Waal's examples of what looks like emotional connection, even altruism, among apes go on and on. For instance, bonobos at the Milwaukee County Zoo were joined by an older male named Kidogo who suffered from a heart condition. New to the building with its complex system of tunnels, this senior ape was confused by the keeper's commands. The other bonobos took him by the hand and led him to where the keepers wanted him to go.

A community of rhesus macaques had a member named Azalea who was developmentally impaired. Rhesus monkeys ordinarily maintain very strict rules of conduct. But, seemingly aware of Azalea's limitations, they gave her a pass for the most irregular behavior, including threatening the alpha male.

Kanzi, a bonobo from the Georgia State University Language Research Center, also in Atlanta, has become famous for his skill in communicating with people. A researcher tried to get Kanzi's younger sister, Tamuli, to respond to certain oral requests, even though she had only very limited exposure to spoken language. Kanzi, a thoughtful big brother, began to act out the meanings for her.

Through accounts like these, most scientists acknowledge what animal lovers have long accepted as common sense: that some of our more intelligent fellow creatures—apes, elephants, porpoises—can be very sensitive to what goes on beneath the skin of others. Just ask any dog or cat owner, and they will tell you that pets know what "their" human is feeling, and know what to do to provide comfort when that human is feeling blue. By the same token, leave your Jack Russell alone too long and you may find that he's taken out his displeasure on the throw pillows from your couch.

Theory of Mind, and Then Some

Among chimps, an aggressor who has attacked and bitten another, but who is now intent on reconciliation, will often look directly at the spot where he injured the other, inspect it, then begin to clean the wound. Bonobos, who, at least in captivity, often have sex face to face, carefully monitor and respond to the expressions and vocalizations of their partner.

The Germans have a word for closely attuned perception of another's emotional state. They call it *Einfühlung*, meaning "feeling into." But can we say that bonobos form an emotional connection when mating? Was Kuni displaying virtue in her attempts to save the starling? Even scientists who share de Waal's views of animal expressiveness are cautious about taking the assignment of human characteristics to other species—anthropomorphism—too far.

I will leave the question of virtue to the philosophers, but in brain science, the word "emotion" carries a fairly dry definition: It is a neural or endocrine response to a stimulus, the function of which is to regulate the organism's inner world in keeping with the outer world of its environment. According to the taxonomy popularized by Antonio Damasio, an emotion is a physical sensation. A "feeling" is an awareness of having an emotion. "Consciousness" is our awareness of the "self" that is having that feeling.

Between Kanzi's impulse to help his sister and the exquisite subtlety of emotion expressed in Shakespeare's sonnets or Molly Bloom's soliloquy lies a fairly broad gulf, and somewhere between the two is where we find the roots of the emotions associated with human loneliness.

We have seen how pleasant physiological sensations motivate us to engage in prosocial behaviors that enhance survival and help perpetuate our genes. We have seen how aversive sensations (loneliness) redirect us away from isolating behaviors that diminish survival and thereby diminish the propagation of our genes. We have also worked our way through a fair number of constituent elements of this pleasure/pain, approach/withdraw system: genetic biases, rewards and punishments from within the social group, the hormones and neurotransmitters that convey the messages that link genes to behaviors and behaviors to genes, and the social feedback loops of co-regulation that complete the circle.

But knowing that birds get around by flying, and that birds have feathers and light bones in order to fly, still does not provide a particularly useful understanding of how birds actually get off the ground. If you want to build a flying machine, it helps to know some aerodynamics. If you want to build more satisfying social connections, it helps to know more about how "emotional connection" occurs in a functional sense, which is to say, how one human brain gains access to the thoughts, feelings, and intentions of another. It also helps to know how and why that system can become overwhelmingly confused.

Theory of mind, which is what we call the ability to have insights into other people's thoughts, feelings, and intentions, develops in humans when we are about two years old. This is the same time when we begin to recognize ourselves in mirrors. So self-awareness and the ability to understand the feelings and intentions signaled by others may be connected. The biologist N. K. Humphrey has even suggested that the adaptive value of being able to detect the emotional state of another person may be what led, not just to the development of human intelligence, but to the development of human consciousness itself.[3]

But beyond our ability to recognize what someone else is experiencing, and to exercise certain discretion in how we respond to it, we have the capacity to spontaneously share the experience. When the young starlet is up at the podium tears streaming down her face, thanking the members of the Academy and her director and her polarity therapist, viewers sitting at home in Bangkok or Birmingham can resonate with her pathos such that they are also moved to cry. But where are the wires or tubes? How does the energy transfer?

Once again, we are up against the question of cause at a distance, and the sympathetic threads that bind us together.

Holding Up the Mirror

Despite our discretionary responses, and no matter where we are on the continuum of social feelings from miserably lonely to "couldn't be happier," our brains respond to other people in ways that are involuntary and automatic. As Adam Smith observed three hundred years ago, we wince when someone else hits his finger.[4] We duck when someone else ducks. We monitor and unconsciously mimic others' eye contact, pauses in speech, and posture. Parents make "yum yum" faces as they try to inspire their baby to imitate—and eat. If one baby cries in the nursery, another baby will pick up the cue. And as mentioned earlier, if you wiggle your foot, chances are I will wiggle my foot even more vigorously in response if I am feeling socially disconnected.

As far as we know, no one has the ability to read another person's mind, but "theory of mind" means that we do not have to sit and ponder, sifting through evidence and making lists, in order to have a pretty good sense of what is going on with other people as we interact with them. The neural basis of this ability begins with the highly refined and unconscious detection and interpretation of movement.

"Action semantics" is the term we apply to this ability to know instantly what someone else's gesture is about, what its goal is, and how it relates to other actions and events. In the increasingly complex social world of even the most primitive humans, this ability allowed our ancestors to make quick and sometimes vitally useful inferences about the intentions of both their allies and their enemies. Studies using fMRI show that certain brain regions—the premotor areas and the inferior frontal gyrus—actually simulate the actions we observe. This simulation, or "action representation," triggers activity in brain regions associated with emotion, such as the insula and the amygdala. This activation tightens the linkage between imitation of another person and identification with that person.[5] Which helps explain all those parents mouthing the lines of dialog from the audience as they watch their third-grader perform in the school play.

Circuits in your brain inhibit you from actually moving while you mentally simulate the actions of others, but meanwhile, your neural experience of the movement creates a template in your brain. This may help explain how children learn to write or to tie their shoes, how young athletes, dancers, or musicians benefit from observing the grand masters, and why millions of people enjoy watching golf on television.

In the 1980s the neurophysiologist Giacomo Rizzolatti began experimenting with macaque monkeys, running electrodes directly into their brains and giving them various objects to handle. The wiring was so precise that it allowed Rizzolatti and his colleagues to identify the specific monkey neurons that were activated at any moment.

When the monkeys carried out an action, such as reaching for a peanut, an area in the premotor cortex called F5 would fire (see Figure 11). But then the scientists noticed something quite unexpected. When one of the researchers picked up a peanut to hand it to the monkey, those same motor neurons in the monkey's brain fired. It was as if the animal itself had picked up the peanut. Likewise, the same neurons that fired when the monkey put a peanut in its mouth would fire when the monkey watched a researcher put a peanut in *his* mouth. "It took us several years to believe what we were seeing," Rizzolatti told the *New York Times* science writer Sandra Blakeslee.[6]

Rizzolatti gave these structures the name "mirror neurons." They fire even when the critical point of the action—the person's hand grasping the peanut, for instance—is hidden from view behind some object, provided that the monkey knows there is a peanut back there. Even simply hearing the action—a peanut shell being cracked—can trigger the response. In all these instances, it is the

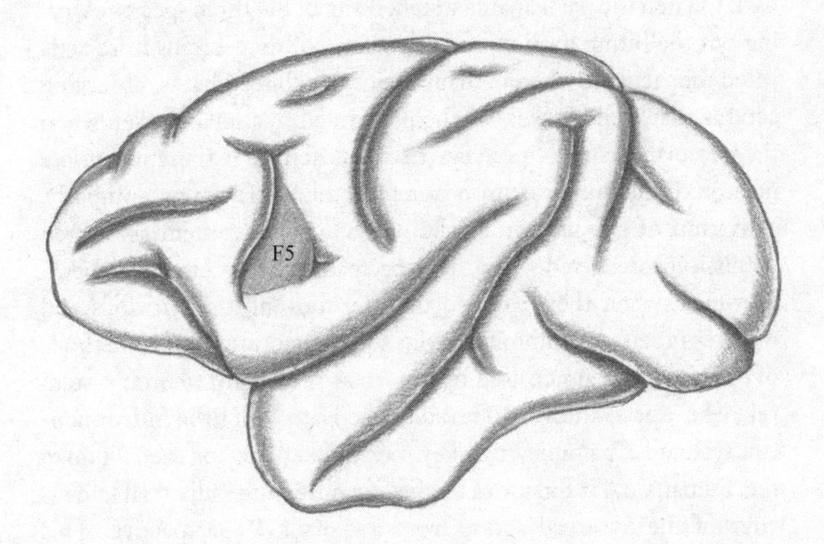

FIGURE 11. The region in the monkey brain (F5) that Giacomo Rizzolatti and colleagues found to contain mirror neurons.

goal rather than the observed action itself that is being mirrored in the monkey's neural response.[7]

To investigate similar "sense and response" mechanisms in humans, Rizzolatti examined the twitching of hand muscles. Working with the neuroscientist Luciano Fadiga, he recorded motor-evoked potentials—the signal that a muscle is about to move—as participants watched an experimenter grasp various objects. The potentials recorded while the participants observed were the same as those recorded when the participants themselves grasped the objects, and as long as the goal of the experimenter was the same (to grasp an object), the potentials were the same whether or not the participants could see the experimenter's hand close around the object.[8]

Rizzolatti and his colleagues confirmed the role of goals in constraining this mental activity by performing brain scans while people watched humans, monkeys, and dogs opening and closing their jaws as if biting. Then they repeated the scans while the study subjects watched humans speak, monkeys smack their lips, and dogs bark.[9] When the participants watched any of the three species carrying out the biting motion, the same areas of their brains were activated that activate when humans themselves bite. That is, observing actions that could reasonably be performed by humans, even when the performers were monkeys or dogs, activated the appropriate portion of the mirror neuron system in the human brain. Similarly, activation of the portion of the mirror neuron system associated with speech occurred when the participants watched humans speak. However, when they watched the oral movements that dogs and monkeys use to communicate—lip smacking and barking, methods of communication not used by humans—the mirror neuron system related to speech movements did not activate. Thus the mirror neuron system isn't simply "monkey see, monkey do," or even "human see, human do." It functions to give the observing individual knowledge of the observed action from a "personal" perspective. This "personal" understanding of others' actions, it appears, promotes our understanding of and resonance with others. It also accounts for the ways in which merely observing can give rise to a sense of shared fate. This neural resonance explains why horror movies can be so horrifying, why we sometimes have the urge to yell out from the audience, "*Don't go in the house!*" It may also explain some of the intense pleasure we derive from watching our loved ones enjoy themselves.

In another study, Rizzolatti and his colleagues used fMRI to scan the brains of fourteen volunteers while they inhaled noxious odors. One of these inhalants was butyric acid, which smells like vomit. They also scanned the brains of the same volunteers as they watched a film of a person smelling the contents of a glass and grimacing in disgust. They found that experiencing a disgusting sensation—smelling the butyric acid—and watching someone else reflect disgust in his facial expression activated the same region of the brain's anterior insula.[10] Emotions such as guilt, embarrassment, and lust are associated with the activation of this same physical structure.

The mind is first about the body, indeed. Yet these "quick and dirty" imitative actions carried out inside our skulls do not involve voluntary control.[11] These imitations and simulations also are faster than other reactions, which means that they can occur even before we are aware of them. Someone attractive smiles at you, and you smile back reflexively, yet it is this automatic response—the physical resonance—that gives you privileged access to the inner experience of another being. You feel his or her flirtation, perhaps become aware of it for the first time—physically—in your own gestures of flirtation.

Assuming your perception is accurate, then, imitation becomes a platform for forming an emotional connection that persists longer than that fleeting moment of resonance. Subtle mimicry can even initiate a positive feedback loop, inasmuch as those being mimicked feel greater rapport with those doing the mimicking.[12] Of course, inaccurate readings can lead to awkwardness, as in: "I'm not smiling at *you* . . . I'm smiling at *him*!"

And as with any intuitive, physical activity, trouble can arise when you "get your head in the way." Thinking too much—what we sometimes call pysching yourself out—can cause a golfer to miss a putt, or a tennis player to hit easy forehands into the net. When conscious awareness, often because of fear, suddenly intrudes into something automatic, such as playing a well-rehearsed piece on the piano, the flow stops and what musicians call muscle memory flies out the window.

Loneliness, of course, is a state of mind that puts your head front and center. By engendering fearful, negative cognitions, it allows the mind to interfere with various forms of resonance that might otherwise flow very naturally into social connection.

Einfühlung

These several varieties of prelinguistic communication—simulation, resonance, motor mimicry—serve as the basis for socially shared representations, as well as for what we call social cognition. They also serve as the basis for automatic coordination and co-regulation among individuals. But they also can form the basis for that deeper social bond whose English name—empathy—comes from the German *Einfühlung* ("feeling into").

Scans from fMRI studies show that several areas of the brain, such as the medial prefrontal cortex, the posterior superior temporal sulcus, and the temporoparietal junction, are activated when we think about other people, or when we try to make sense of social relationships (see Figure 12).[13] In one study, scientists observed that the greater the degree of activation in a person's posterior superior temporal sulcus, the greater her likelihood of behaving altruistically.[14] This part of the brain plays a leading role in the perception of agency, and, as a result, it can be involved in integrating our personal experience into a meaningful narrative. Thus it appears that altruism is grounded in our "thinking about" and in our "trying to understand" our life with other people. Put another way, altruism emerges from the kind of narrative we construct about human agency or responsibility.

The processing of information that leads to our seeing the world in this seamless way relies on a coordinated effort involving several other brain regions as well: those specialized for dealing with

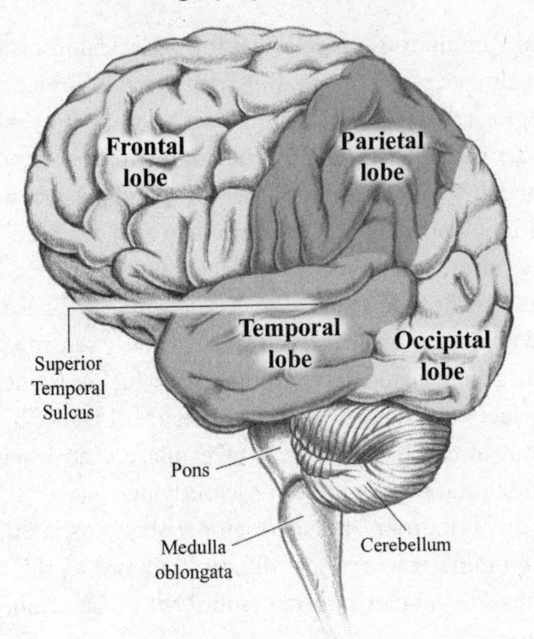

FIGURE 12.

Right and left frontal lobes	control executive functions such as thinking, planning, problem solving, social intelligence, and impulse control
Parietal lobes	integrate information from the senses and control relational and associative functions such as visuospatial processing
Occipital lobes	early processing of visual information
Temporal lobes	memory, auditory processing, communication, biological motion perception (especially along the superior temporal sulcus), and other forms of high-level visual processing of, for instance, faces and scenes
Pons	relays sensory information between the cerebellum and the higher areas of the brain, controls waking and arousal, regulates the muscles of facial expressions, and is involved in dreaming
Medulla oblongata	controls basic autonomic functions such as respiration, heartbeat, and blood pressure
Cerebellum	contributes to the integration of sensory and motor processes, thereby permitting motor coordination, precise movements, motor learning, as well as our ability to walk smoothly and in balance

emotional stimuli, those specialized for nonsocial information, and those largely specialized for dealing with social information.[15] These neural regions are widely distributed across the brain, and different ones are activated depending on the context. For instance, perception of faces involves a region of the lower and posterior part of the temporal lobe known as the fusiform face area, or FFA.[16] Recognition of the emotions conveyed by a face, however, depends on other structures that decode specific emotional signals. These structures include the anterior insula, which is particularly sensitive to expressions of disgust or pain, and the amygdala, which is particularly sensitive to faces expressing fear (see Figure 13).[17] This sensitive recognition of emotional content takes place even when the faces are presented too rapidly to be consciously perceived.[18]

What this brief tour of neuroanatomy shows us is the degree to which the brain's resources are allocated to two of the things that matter most for human survival: emotional recognition, and other human beings.[19]

The superior temporal sulcus, the medial prefrontal cortex, and the amygdala are all sensitive to the emotional content of pictures as well as to the social content.[20] However, these brain regions deal with emotional timbre (happy or sad) as a separate issue from whether the picture portrays people or objects. This has allowed us to determine that, regardless of emotional timbre, stimuli (such as pictures) that depict people typically evoke greater brain activation than those depicting objects even when the luminance or novelty or other features are comparable.[21] A picture of a sad clown, in other words, produces more activity in the brain, and in more different brain regions, than does a picture of a gloomy forest.

This disproportionate allocation of resources is consistent with the "social brain hypothesis," which holds that it was the complexities of social living that drove the rapid expansion of the human cortex.[22] A spider monkey's brain is perfectly adequate for the challenges of finding enough to eat while not stepping on a snake. It is also adequate for getting by in a small troop with rigid social rules. But greater behavioral latitude means greater social complexity. The more demanding mental challenges involve sorting out friend

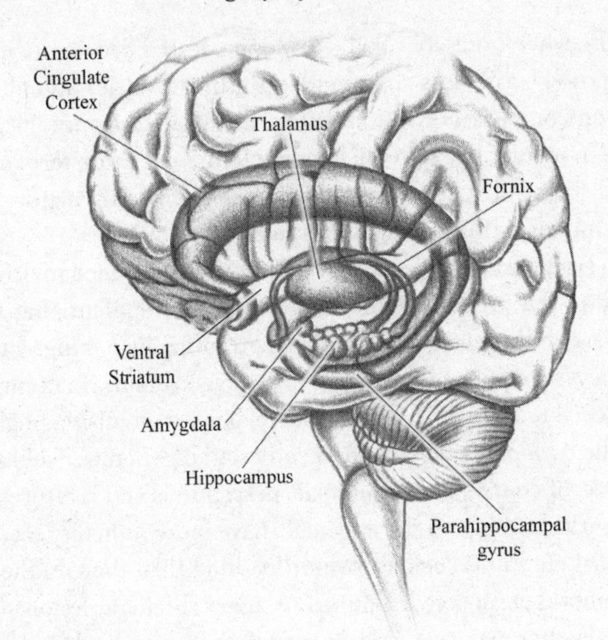

FIGURE 13. Here we look "through" the brain's outer layers if they were translucent, to see interior structures.

Anterior cingulate cortex	involved in processes such as error detection, modulation of attention, and executive processes
Thalamus	relays sensory information and, via connections to the frontal lobe, modulates levels of awareness
Amygdala	involved in social cognition, emotional learning, and memory consolidation; as well contributes an initial motivational tendency to attend and approach or avoid
Ventral striatum	involved in rewarding feelings, planning and modulating movement, and processing novel or intense stimuli
Hippocampus	involved in spatial navigation and in the formation of new memories that can be verbalized
Parahippocampal gyrus	involved in high-level visual processing of scenes and in aspects of memory
Fusiform gyrus	(not visible here; located below the parahippocampal gyrus) serves in the high-level visual processing of faces

from foe when both are capable of sophisticated deception, negotiating power structures that include shifting rivalries and alliances based on complex motivations, using language to communicate (as well as to manipulate others), juggling long- and short-term mating relationships not rigidly constrained by the female's ovulatory cycle, and grappling with ever-changing cultural evolution.[23]

The last point is the rocket propellant that continued to drive and accelerate the process of human mental development. Big brains don't rest on their laurels—they keep inventing new things and creating new situations that demand even bigger brains. Inattention to exact social meaning can cause trouble, but so can misreading social cues, the twin perils that so often confound us when we feel lonely.

Injury, of course, can also impair perceptions and interpretation. People with damage to the amygdala have more difficulty recognizing social emotions such as love or loathing than they do the most basic emotions such as happiness or anger. Bilateral lesions of the amygdala alter eye gaze, and because fear is expressed in the eyes, patients with bilateral lesions in the amygdala do not accurately perceive others' fear, and thus they have a hard time assessing trustworthiness.[24] Patients with amygdala lesions can identify positive and negative stimuli, but even though they rate positive stimuli (a giggle) as worthy of special attention, just as you or I would, they rate negative stimuli (a growl) as no more arousing than a sound with neutral implications.[25] And yet, even when participants in experiments are directed to concentrate on a nonsocial consideration such as whether a picture is pleasant or unpleasant, their brains continue to focus on whether or not there are people in the picture.[26] This happens automatically because, once again, our big brains did not evolve in order to evaluate art or to solve quadratic equations. They evolved because it was to our adaptive advantage to be able to process and manage complex and dynamic *social* information.

Our ability to form impressions of other people, including our ability to adopt another person's perspective and assign mental states and intentions to them (theory of mind), derives from a different set of structures, including the medial prefrontal cortex, the anterior cingulate cortex, and the temporoparietal junction.[27]

But the ability to assign mental states and intentions still does not guarantee accuracy. Perceptions of others' mental states are, like our experiences of empathy, very much a matter of the narrative we construct around them, and these interpretations are easily distorted by the one pain so devastating that it disrupts our executive function—loneliness.

Watching Closely

In a study of how people monitor social cues, when researchers gave participants facts related to interpersonal or collective social ties presented in a diary format, those who were lonely remembered a greater proportion of this information than did those who were not lonely. Feeling lonely increases a person's attentiveness to social cues just as being hungry increases a person's attentiveness to food cues.[28]

The same researchers then went on to test how skilled the lonely were in decoding and inferring meaning from less explicit, nonverbal modes of expression. They presented images of twenty-four male and female faces depicting four emotions—anger, fear, happiness, and sadness—in two modes, high intensity and low intensity. The faces appeared individually for only one second, during which participants had to judge the emotional timbre. The higher the participants' level of loneliness, the less accurate their interpretation of the facial expressions.[29]

In another study the researchers asked three groups of participants to "relive" one of three different experiences by writing about them: a time in which they had felt intensely rejected, a time in which they had felt intense failure at a nonsocial, intellectual challenge, or simply the experience of their walk or drive to the campus that morning. Researchers then tested all three groups for their ability to perceive subtleties in spoken language. People who had just relived rejection experiences showed greater attention to vocal tone, but they too were less accurate in interpreting specific meaning.[30] All of which helps account for the feeling of threat we often associate with social settings when we feel lonely. Walking into a party, a classroom, or a business meeting, our internal monologue when we feel lonely is often about our fear of negative evaluation. What are these people going to think of me? I can't believe I wore this outfit! I don't know a soul—they'll think I'm a loser.

But variations in response to social cues revealed by fMRI scans also show us why loneliness deprives us of some of the pleasure we might otherwise find in the connections we have. In the Brain Imaging Center at the University of Chicago, we asked a group of volunteers to look at photographs while undergoing fMRI.[31] The pictures we showed them were of objects or of people, and many were selected to have an emotional impact that was once again positive or negative. We made sure that the degree of positive or negative impact was rated consistently for both kinds of photographs. A picture of an object rated "highly negative" (a disgustingly filthy toilet), for instance, had to be equal in negative impact to a picture of a person rated "highly negative" (a man who had been beaten and bloodied). Following the scan and the viewing of pictures, we measured each of the participants' level of loneliness.

As indicated in Figure 14, nonlonely participants showed greater activity in the ventral striatum, one of the brain's "reward centers,"

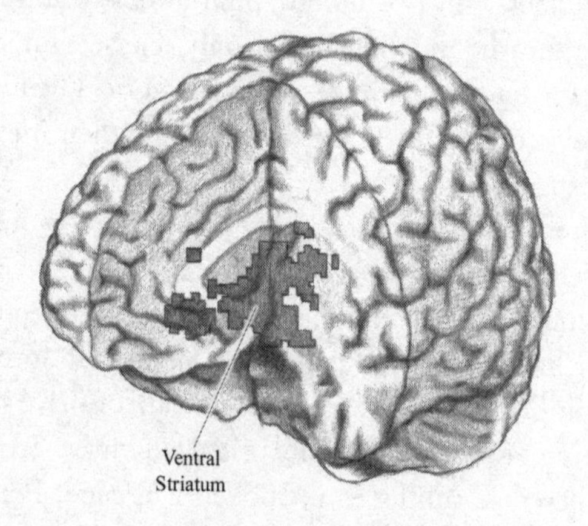

FIGURE 14. People who were not lonely, relative to those who were lonely, showed more activity in a large limbic region including the ventral striatum (a reward area of the brain) than lonely individuals when looking at pleasant pictures of people than when looking at equally pleasant pictures of objects.

when they saw a pleasant image of a person (a smiling farmer) than when they saw an equally pleasant picture of an object (a flower arrangement). For the nonlonely, a positive image of another human being obviously meant something special—it gave them a specific emotional boost as evidenced by changes in this specific area that registers pleasure. Lonely participants, however, when they viewed positive images of people, did not register the same boost: The activation of the ventral striatum in response to a happy face was in fact a bit weaker than when they saw the pleasant picture of the flowers. This finding aligns with self-reports in which lonely individuals said that they found positive social interactions to be less of an uplift than did their nonlonely counterparts.[32]

In the same study, when the photographs were negative, the patterns of brain activation were equally revealing about the experience of loneliness, moment to moment (see Figure 15). Whereas nonlonely participants paid comparable attention to negative images of

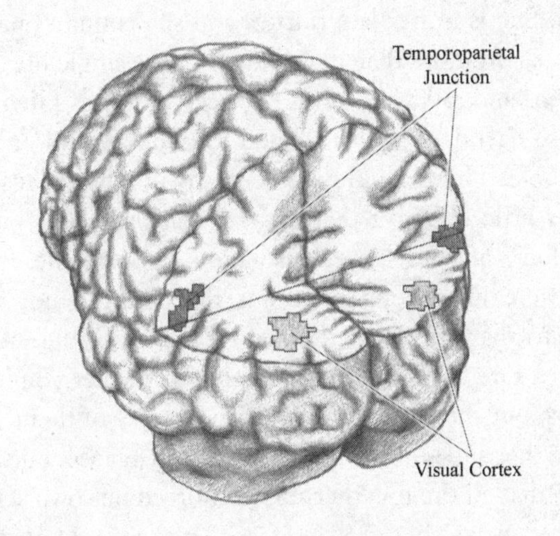

FIGURE 15. People who were lonely showed more activity in the visual cortex—and less in the temporoparietal junction—when looking at unpleasant pictures of people than when looking at equally unpleasant pictures of objects.

people and objects, lonely individuals paid far more attention to the negative images depicting people. In both cases, the measure was activation of their visual cortex.

When we observed the temporoparietal junction, an area involved in theory of mind and in perspective taking, the pattern of activation was the opposite. Lonely individuals appeared more likely than their nonlonely counterparts to respond to pictures of people in peril from their first-person perspective (as indicated by weaker activation of the temporoparietal junction). When we feel lonely, we tend to scan the horizon for any possibility of social danger, but with an eye toward protecting ourselves rather than with genuine concern for what others may be thinking or feeling.

When you combine less pleasure derived from pleasant company and a narrowed perspective inordinately focused on threats, real or imagined, in social situations, the unfortunate result is less socially skilled responses that can serve to reinforce the lonely person's isolation.

I was at a wedding not long ago that was the quintessential joyful occasion, except for one woman who seemed unable to share in the warmth. She is in her late thirties and she is quite open about the fact that she is more than a little tired of the single life and ready to find a husband and have kids. The reception was filled with family and close friends, but she seemed to glide around encased in her own bubble. There could have been any number of reasons for her to feel a little out of it that day—health problems, job problems, depression—but her sense of isolation was palpable. It wasn't that others were unwelcoming; it was that she did not appear fully "there" to share in the emotional resonance of the moment. This was also a rare chance for her to get to know her young nieces and nephews, but she did not really attend to any of them. It may have been that her shoes did not fit properly and her feet hurt, but it's also possible that all the togetherness reinforced her own sense of being left out, as well as her growing fear that she would never have children of her own. The distraction and dysregulation caused by lonely feelings may have kept her from enjoying the connection offered by dozens of warm friends and family members, and also from appreciating that these children were "hers" in a different sense: children for whom she could be a loving aunt. Her conversations were brief and a bit strained, and she left early.

The inability to relish the positives has further implications for those who carry a subjective sense of isolation into their intimate relationships. "Social capitalization" is the term psychologists use to describe the support and reinforcement a person receives from his or her partner after a *positive* experience. Studies show that truly enjoying these positives and making the most of them is even more important to the health of a marriage or other intimate relationship than being supportive during hard times. Sharing the joy in your partner's promotion, it seems, actually can be more important than being attentive when she gets passed over. Similarly, another study showed that when it comes to problem solving within a marriage, remaining cheerful and pleasant in outlook—even when that cheerfulness is combined with less than perfect communication skills—was far more predictive of keeping your partner happy than was being a grump who somehow manages to do or say exactly the right thing.[33]

When Loneliness Interferes with Einfühlung

If you show signs of suffering, I may be able to see what you are going through and I may feel bad about it, but that is not yet empathy. I could be a narcissist responding with, "What a buzz-kill . . . you're bringing me down." I could be a kindly but rather detached therapist simply revisiting your last phrase as a prompt—"You say you're feeling distressed because you've lost your boyfriend"—all the while wondering how London gold futures held at the close of trading.

Many of us have experienced the ambient stress that can take over a household when one spouse comes home panicked about work and the traffic and the price of gasoline, and within minutes everyone is feeling it, right down to the baby and the dog and the cat. Similarly, a shared motor response alone can create what is called "emotional contagion," but not empathy. The way to avoid this kind of second-hand stress or other emotional overflow is not as simple as avoiding secondhand smoke. We don't want to simply avoid—and thus become inured to—another's pain. The better idea would be to increase our own emotional self-regulation so that we can respond appropriately.

My colleague Jean Decety studies the neuroscience of empathy, and he has identified four essential elements in this form of connection: shared affect, awareness that the other is separate from the self, the mental flexibility nonetheless to "put yourself into the other's shoes," and the emotional self-regulation necessary to produce an appropriate response.

Decety has been able to identify these four separate elements and to isolate them in different areas of the brain. He has determined that, with the exception of shared affect, each of the elements of empathy requires executive function. When loneliness takes over, then, the feelings of isolation that contribute to disrupting executive control and self-regulation can interfere with truly empathic responses as well.

Under experimental conditions, Decety showed pairs of photographs to sixty-four test subjects. One photo would depict an ordinary situation, say a hand with clippers pruning a twig; the other would convey pain, perhaps the same hand squeezed in the clipper's blades. One picture would show a bare foot beside an opening door; the other would show the door opening onto the foot. When the pictures switched from ordinary to awful, the subject's anterior insula and anterior cingulate cortex would light up. The dorsal anterior cingulate cortex is responsible for coordinating motivation and affect in response to pain—an executive control function that allows us to respond to negative events in ways that are measured and appropriate. In Decety's studies, the more painful the experience being depicted, the more intensely these brain regions were activated.[34]

But just as we don't want to be inured to suffering, we also don't want to surrender to "empathic overarousal." We will not be of much use helping our friend reach the emergency room or calming our screaming child if we are sympathetically writhing in agony to the same degree that he or she is. The brain regions that Decety and his colleagues have shown to be involved in distinguishing between self-produced actions and those generated by others are the medial prefrontal cortex, the temporoparietal junction, and the dorsal anterior cingulate. These regions, coupled with the anterior cingulate and lateral prefrontal cortex, which are involved in regulating emotions, permit an appropriately measured empathic response to seeing another person in pain.[35]

Just getting through an average day requires staying on a fairly even keel, which, again, requires emotional self-regulation.[36] Loneliness uniquely can cause us to overshoot, as well as undershoot, the happy medium of well-regulated emotional balance. Making matters even more difficult, when we feel lonely we feel less of the uplifts that most people feel simply from seeing others in happy circumstances.

The Illusion of Me over Here

Social neuroscience shows us not only that there is no magical boundary between mind and body, but that the boundaries we have always assumed to exist between ourselves and others are not nearly as fixed as we once imagined.

The social psychologist Gün Semin argues for the existence of what he calls "co-cognition." The brain's way of forming representations means that two or five or fifty people can roughly share the same perspective. That cognitive sharing is what lends extra excitement to listening to jazz or to a jam band, to watching improvisational comedy, or to seeing teammates on the field moving the ball, anticipating what each needs to do to get the opposition off balance and the ball into scoring position. To some degree we experience what all the other spectators experience; depending on our level of skill and engagement, we also experience to some degree what the participants experience. This sharing also permits members of cardiac surgery teams and World Cup soccer teams to anticipate one another's wishes and needs and to problem solve and respond at a speed that would not be possible for a single individual. We are transported into intense coordination and synchronization that move at a pace that can be quicker than conscious thought. Through "cognitive sharing" we briefly transcend the boundaries of the self.

Coaches, corporate leaders, and motivational speakers are fond of telling us that if we can imagine it, we can achieve it. Co-cognition is one reason why corporations, governments, and other large organizations make enormous efforts to get everyone trying to imagine the same thing, focusing their mental energy on a specific, highly refined "mission statement." At the other end of the cultural spectrum, the same fundamental idea—reduced to the cliché of "we are all one"—is central to various forms of mysticism. The Zen master Yasutani Roshi expresses it this way: "The fundamental delusion of humanity is to suppose that I am here and you are out there."[37]

As we will explore in the chapters ahead, the abilities to see the bigger picture and to creatively and often collectively adapt across contexts are perhaps our most distinctive human attributes. But sharing a vision with others is not a simple matter of having matching neural structures. It is hard enough to have a consistent mission statement in our own minds. "The human heart in conflict with itself" is the only thing worth writing about, said William Faulkner, and from *Oedipus Rex* to *ER*, human social structures, as well as human anatomy and physiology, have given novelists and playwrights ample material. Anyone can have debilitating inner conflicts. It's just that when we feel lonely, we are likely to experience far more than our share.

CHAPTER TEN

conflicted by nature

Part of the process of maturing into adulthood is gaining control over our unbridled emotions and impulses. Loneliness diminishes that control, then causes more trouble as it engenders other negative emotions such as hostility and anxiety. When we feel lonely, people may see us as aloof, less than empathic, socially insensitive, perhaps even ungenerous, when, deep down, what's really going on is that our cognition and self-regulation are being distorted by fear. But no matter how socially contented we are, none of us leaves our less controlled responses entirely behind.

The great neurologist John Hughlings Jackson was the first to recognize that during maturation, individual development follows the same general pattern of layered upgrades, rather than downloads and overwrites, that unfolded over the course of the brain's evolution. This means that as we mature, rather than dispensing with our more infantile and animalistic impulses, we merely bring more sophisticated forms of processing online, which give us the ability to inhibit—sometimes only with considerable effort—those lower-level responses. This already complicated arrangement is further complicated by the fact that our neural wiring does not contain a simple, binary switch for good sensations and bad sensations labeled "pleasure" and "pain." Instead, these sensations come in multiple varieties, and evolution sculpted them to operate as carrots and sticks both separately and in coordination across the many different layers of the nervous system.

Layering higher levels of function on top of lower levels allows us to make use of stimulus-response mechanisms from the spinal cord, brain stem, and limbic region when we need them: "quick and dirty" reflexive responses such as "baby falling—reach out to catch!" But we also have the more sophisticated cortical functions that allow for prolonged consideration, mental time travel, and nuanced decisionmaking.

Despite its advantages, maintaining the low road as well as the high road can sometimes be a prescription for anguish, ambivalence, and being at cross-purposes with ourselves. Loneliness, as we have seen, is a great enabler of such conflicts, causing us to seek warmth and companionship while at the same time allowing fearful perceptions to make us harsh and critical toward those we wish to be near.

Holding Your Horses

Plato saw human nature as a charioteer trying to control two horses, one representing our "noble" impulses, the other our unruly passions. But that dual responsibility would be child's play compared with the complexity that neuroscience now shows us actually to be the case. The multiple neural pathways in our brains are not divided along simple lines such as good/bad, noble/base, logical/passionate; in fact, they are not even arranged in a typical hierarchy or a consistent chain of command. Instead, they are organized into a complex amalgam that the neuroscientist Gary Berntson has characterized as a "heterarchy."

While the higher capabilities of the frontal cortex exercise executive control, the limbic region, or midbrain, serves as a processing platform for information and regulation. It takes in sensory information, transmits it up the chain of command, and then conveys the messages back down the line in order to carry out our intentions.

But as we have seen again and again, both high road and low road are subject to the influences of social context, including whether we feel warmly included or distressingly alone.

The prefrontal cortex that is central to rational planning and deliberate execution of behavior is also critically involved in the regulation of emotion. When people are asked to reflect on themselves or on others, the prefrontal cortex is where we see heightened activation on brain scans.[1] So this newest part of the brain is a charioteer with multiple reins. Higher-order control (working memory, attention, choice and decisionmaking) has the challenge of imposing order on lower-order processes such as affect, drive, and motivation. For instance, we have a reflexive tendency to spit out bitter substances. This response developed because the poisons we once encountered in our natural environment tended to taste bitter. Some cough syrups also taste bitter, though, and here is where a well-functioning frontal cortex comes into play. A child may cry and gag when given such remedies, but with maturity we learn to override those natural impulses and "take our medicine."

The distributed arrangement of neural processing in the human brain also has the great advantage of allowing increased behavioral flexibility and contextual control. But we need the executive brain to filter out extraneous thoughts, focus our minds, and regulate our more deeply embedded, sometimes primitive, responses. And here again, loneliness gets in the way.

With the dichotic listening task described in Chapter Three, we gave participants conflicting auditory signals. In 1935 the psychologist John Ridley Stroop developed a way to measure conflicting signals that are cognitive. Psychologists, administering what became known as the Stroop Test, show participants a list of color names on a page, but the word "red" will be written in yellow or green, the word "yellow" will be written in blue or red, and so on. Then they ask participants to name the colors. The dissonance between the visual information (the color itself) and the verbal information (the word written out) interferes, causing a tiny delay as the participant tries to make sense of the competing stimuli.

We set up a test of interference based on the Stroop model by printing out various words on a page in many different colors.[2] Among those random choices were some emotion words, such as "fear," and social words, such as "compete." The task was to name the color in which each word was printed. For the social words, participants who were lonely took a split second longer than those who were nonlonely to identify the colors. The delay indicated an interference effect. Even when the task had nothing to do with sociality, and with no awareness of any intention to do so, the lonely participants were scanning for, and being distracted by, social information. Social words associated with negative emotions, such as "torture," ramped up the effect even more.

Just as dieters, despite their best efforts, find themselves transfixed by food, the lonely, far more than others, are focused on social connection and social rejection in everything they see and do. And thus even everyday social situations can do for the lonely what the sight of all that candy did for the mathematically adept chimp, Sheba, which was to trump her knowledge of how to play the game.

The Stories We Tell

A basketball-playing friend of mine used to wander into pickup games while traveling on business. Charlie looks the part—he's tall and lanky—and in fact he's pretty good. He was even on some championship teams in school, but he was always just a rebounder and a role player, never a shooter, and certainly never the "go-to guy." But on one particular day, in an unfamiliar gym in a city far from Charlie's home, the only thing the other players saw walking onto the court was a new face with some height. The men sorted themselves into teams and began to play, and the first time Charlie got his hands on the ball, he happened to be open just inside the top of the circle, so he took the shot and made it—nothing but net. One or two of his teammates gave him a nod, but no big deal. On their second possession, Charlie once again got the ball in good position, so once again he took the shot, with another beautiful arc that ripped right through the net. More nods this time, combined with a couple of smiles and a high five, but Charlie remained nonchalant—at least in the eyes of the other players. He tried to make it look as if he did this all the time. After all, nobody else knew anything about his twenty-year history of performance anxiety, inevitably passing off to avoid throwing up clunkers under pressure.

His teammates were feeling good now—obviously they had a scoring machine on board. And so it went for the next ten minutes. They fed Charlie the ball, he continued to shoot, and he went eight for eight from the outside, plus one spectacular drive to the basket. His team smoked the other guys, and then with a wave of his hand, Charlie said, "Thanks . . . gotta go," and ducked into the locker room. Charlie's deep, dark secret was that he was only *impersonating* a great shooter, and that it was time for him to quit while he was ahead. He had never shot like that in his life, but the luck of the first attempt had given him confidence for the second, which was then reinforced by his complete anonymity. For all the others knew, he was a small-time Kobe Bryant. So, for that one brief scrimmage, he was.

Biased meaning-making is a powerful force that can help us reach new heights or keep us from getting out of bed in the morning. And, as we have seen in many other contexts, such an effect is not "all in our heads." Whenever we fear that we might fail at an important task, this bias can cause us to handicap ourselves, producing insurmountable obstacles to our own success. But loneliness, and the egocentrism it generates, can turn this natural tendency into a serious and persistent state of affairs. Even when the important task is to achieve human connection, seeing ourselves as congenital outsiders, subject to threats, hungry and needing to be fed, undermines our best efforts. Then again, it is this same human ability to be the "architects of our own reality" that gives us the key we need to emerge from our solitary confinement.

When We Get It Wrong

As we try to determine the meaning of events around us, we humans are not particularly good at knowing the causes of our own feelings or behavior. We overestimate our own strengths and underestimate our faults. We overestimate the importance of our contribution to group activities, the pervasiveness of our beliefs within the wider population, and the likelihood that an event we desire will occur.[3] At the same time we underestimate the contribution of others, as well as the likelihood that risks in the world apply to us. Events that unfold unexpectedly are not reasoned about as much as they are rationalized, and the act of remembering itself—even the "eyewitness testimony" offered in courtrooms—is far more of a biased reconstruction than an accurate recollection of events.[4] Subtle reminders of mortality can push people to blame the victim. Female jurors are actually more likely than their male counterparts to believe that a rape victim somehow contributed to her fate. "After all," the juror thinks, "if this happened to *her* without her behaving badly or taking stupid risks, then it could happen to *me*! She has to be somewhat responsible for what happened; otherwise, I can never feel safe." We are also very poor judges of how long particular experiences will make us feel either good or bad. In virtually every domain we confirm what we already believed to be true. We say that opposites attract with the same certainty we express when we say that birds of a feather flock together. Or, as the sage of the New York Yankees Casey Stengel put it, "Good pitching will always stop good hitting, and vice versa."

Amid all the standard distortions we engage in, as well as the kind of interference effects we saw with Sheba and the candy, loneliness also sets us apart by making us more fragile, negative, and self-critical. In one study participants performed a simple task, after which they received feedback evaluating their success or failure. The higher an individual's loneliness, the more likely she was to attribute failure to something about herself and success to something about the situation.[5] For the nonlonely public at large, it is far more the norm to see bad luck in one's failures and to take personal credit for success, even when it comes on a lucky break.[6]

One of the distinguishing characteristics of people who have become chronically lonely is the perception that they are doomed to social failure, with little if any control over external circumstances.

Awash in pessimism, and feeling the need to protect themselves at every turn, they tend to withdraw, or to rely on the passive forms of coping under stress that elevate their total peripheral resistance and, eventually, their blood pressure.[7] The social strategy that loneliness induces—high in social avoidance, low in social approach—also predicts future loneliness. The cynical worldview induced by loneliness, which consists of alienation and little faith in others, in turn, has been shown to contribute to actual social rejection. This is how feeling lonely creates self-fulfilling prophesies. If you maintain a subjective sense of rejection long enough, over time you are far more likely to confront the actual social rejection that you dread.[8]

This process was demonstrated in another iterated prisoner's dilemma game in which study participants, some of them lonely, some nonlonely, played against a person they didn't know.[9] In this version of the game, the participants played for money. Before each trial, players told their opponents whether they intended to act on the basis of loyalty or betrayal—but the opponents did not know if they were lying or telling the truth. If one player followed through on his stated intention but the other did not, the double-cross would lead to the trusting player's losing the game. But this being a study and not just a parlor game, the opponent was, in fact, a researcher who pretended to be a study participant and who always responded the same way the real participant had on the prior trial—that is, the researcher used a tit-for-tat strategy. Because the researcher always copied the previous move, the actual participants were determining how the events unfolded, even though they did not realize this to be the case. During early trials, lonely and nonlonely individuals were equally cooperative. As play continued, however, and occasional defections occurred by players only to be followed by defections by their opponent, the lonely players became much less trusting. Their interactions devolved into consistent defection and acrimony. Meanwhile the nonlonely players, despite occasional defections by themselves and by the researcher, were generally cooperative throughout the game. The different social realities created by the lonely and nonlonely participants reflected their different default expectations about the nature of others.

The eminent American psychiatrist Harry Stack Sullivan described loneliness as an experience "so terrible that it practically fables clear recall." For young people especially, he said, the fear of ostracism is "the fear of being accepted by no one of those whom one must have as models for learning how to be human."[10]

Seen in those terms, it is no wonder that loneliness evokes such feelings of dread, or that the young are often so desperate to connect with peers that they sacrifice their own identity as well as their good judgment. The fear of being excluded can make anyone, young or old, do foolish things, including self-defeating things. In an effort to protect themselves against disappointment and the pain of rejection, the lonely can come up with endless numbers of reasons why a particular effort to reach out will be pointless, or why a particular relationship will never work. This may help explain why, when we're feeling lonely, we undermine ourselves by assuming that we lack social skills that in fact, we do have available.

Mind over Matter

Although he put the words into the mouth of Satan, John Milton summed up much of the human condition when he wrote:

> *The mind is its own place, and in itself*
> *Can make a Heav'n of Hell, a Hell of Heav'n.*

Shakespeare's variation on the theme was to say: "There is nothing either good or bad but thinking makes it so."[11]

Human beings are inherently meaning-making creatures, and the lonely are hardly unique in interpreting social cues through a highly subjective lens. The human brain must take disparate, atomistic snips of sensory input and weave them all into a "theory of the case," an interpretation of time and space, cause and effect, that allows us to survive today, plan for tomorrow, and make sense of the past. Ideally, the narrative we construct aligns with objective reality well enough for us to appropriately address the problems confronting us in the real world. However, there is no guarantee. Reflexively, the hypersocial human brain registers three dots in a triangular pattern as representing a human face, but sometime a pattern of three dots is merely three dots.

In the 1940s the psychologists Fritz Heider and Mary-Ann Simmel produced a brief animated film portraying a small triangle, a small circle, and a large triangle that moved around and into a large rectangle. The film was nothing more than these swirling geometric shapes, yet everyone who viewed it "saw" a social drama unfold, complete with intentions, plans, and an emotional subtext. This is simply the human brain doing what it does best—constructing a "reality" out of whatever sensory data it receives.[12]

In the same way that observers could find a story in moving geometric shapes, young children, before they develop theory of mind, promiscuously project their own thoughts and experiences onto other people. During the infancy of the human race, it was this same tendency, most scholars agree, that gave rise to early religions, in which natural forces were given names and complex personal histories—humanlike attributes that served as source material for the first myths and legends.

From Plato's charioteer to Freud's tortured subconscious, philosophers have seen a rational and admirable side of human nature combined with a darker, emotional side. But social neuroscience leads toward a more unified view. Because the emotional system that governs human self-preservation was built for a primitive environment and simple, direct dangers, it can be extremely naïve. It is impressionable and prefers shallow, social, and anecdotal information to abstract data. But the same irrational processes that can bring us down can also be the foundation of our finest qualities as human beings.

Hope entails irrationality. Positive illusions about one's spouse contribute to longer and happier marriages.[13] Without an optimistically biased weighing of the odds, few people would start new ventures. Going by the statistics alone, it is irrational for any individual to assume that he or she can start a successful business, paint a canvas that will sell to a serious collector, write a novel worth reading, make a significant contribution in science, or marry for life.

Bias also can result from the simple need to take cognitive shortcuts. Confronted with more information than we can possibly process, we tend to economize on thought when forming beliefs that are not immediate to our survival: beliefs about politics, culture, or religion. At other times we make choices while remaining completely unaware of the embedded images, preconceptions, and prejudices that govern our preferences. But it is the confluence of the rational and the emotional/irrational that determines much of the narrative of our lives. The same experience can be a challenge or a nightmare depending on how we frame it, the same glass half empty or half full. And it is this threat-surveillance system, coupled with hyperattention to social information—social information often distorted by a defensively egocentric perspective—that we need help escaping once we have slipped into a period of prolonged loneliness. This negative framing is the Catch-22 that makes people with a heartfelt and deeply rooted need for social connection wind up busily creating the very roadblocks that will frustrate that need. And that Catch-22 can ensnare any one of us at any stage of life.

A teenager walks into a party at his new high school, a twenty-something shows up for her first day at a new job knowing no one, an elderly widow attends an event at a friend's church or club not long after the death of her spouse. A sense of isolation can make any of them feel unsafe. When we feel unsafe, we do the same thing a hunter-gatherer on the plains of Africa would do—we scan the horizon for threats. And just like a hunter-gatherer hearing an ominous sound in the brush, the lonely person too often assumes the worst, tightens up, and goes into the psychological equivalent of a protective crouch.

It is hard for most of us to be articulate about our emotions under the best of circumstances. It is that much harder when we have intense sensations of threat flooding our body with stress hormones, and no conscious awareness of what is causing us to sweat or to take rapid, shallow breaths. Accordingly, a great many of us spend a great portion of our lives acting a bit like agitated wind-up dolls, walking into the same walls again and again, wondering why we are trapped inside such a small, lonely room—a room that we ourselves have inadvertently helped design.

The Realities We Construct

Happily, the same cognitive capacity allows us to become conscious of what confines us, and to design doors and windows that open wide. But again, those liberating social cognitions don't come with just a snap of the fingers.

The kind of "reality" we construct for ourselves also determines in large part how others view us and act toward us. They "see" the reality we construct, use it to define us, then act toward us on the basis of that assessment. That's why gaining freedom from loneliness requires a bit of retraining, and a bit of discipline—because the mind's tendency to twist reality into shapes unrecognizable to others is nothing transitory or superficial.

When we feel lonely, we are painfully aware that our social needs are not being met; at the same time, we have a greater tendency to see ourselves as having little control over our ability to fulfill those needs.[14] The prejudiced opinions of others always play a role in this negative feedback loop. If people expect a new acquaintance to be fun and nice, they will behave in a fashion that draws out the pleasant and enjoyable side of that new acquaintance. If parents or teachers think a child is intelligent, they will do and say things that will encourage that child to exercise her intelligence. In one study, participants were introduced to opposite-gender partners after being told that the person they were about to meet was either lonely or not lonely. They subsequently rated the partners they had been primed to consider lonely as less sociable than the others. They also behaved in a less sociable manner toward the partners they expected to be lonely.[15]

When our negative social expectations elicit behaviors from others that validate our fears, the experience makes us even more likely to behave in self-protective ways that spin the feedback loop further and faster toward even more isolation.[16]

So while any of us may become lonely because of a genetic disposition coupled with an unfortunate situation, we remain lonely partly because of the manner in which we and others think. As the trap of loneliness becomes more a function of social expectations and aspirations, the literal reality recedes in importance.

One might expect that a lonely person, hungry to fulfill unmet social needs, would be very accepting of a new acquaintance, just as a famished person might take pleasure in food that was not perfectly prepared or her favorite item on the menu. However, when people feel lonely they are actually far less accepting of potential new friends than when they feel socially contented.[17] Studies show that lonely undergraduates hold more negative perceptions of their roommates than do their nonlonely peers. This divide between the lonely and the nonlonely in their perceptions was even larger when the others being perceived were their suite mates, was larger still for floor mates, and was even more pronounced for students on other floors of their dormitories.[18]

Time also plays a role in constructing these negative "realities." Researchers asked participants to interact with a friend, and immediately thereafter to rate the quality of the relationship and the quality of the communication. Participants then watched a videotape of the same social exchange and rated it again. A few weeks later the researchers reminded participants of their earlier exchange with their friend and asked them once again to rate the quality of interaction and communication. The participants watched the videotape once more and, once more, rated the interaction. At all four measurement points, lonely individuals rated relationship quality more negatively than did nonlonely individuals. But the further in time they were removed from the social exchange, the more negatively they rated it. They were especially negative after each viewing of the videotape.[19] When they rated the interaction soon after it happened, it appears that their negative social cognition was reined in by their understanding of the reasons for their friend's behavior. As time went on and memory for the underlying subtext faded, however, the constraints faded as well. The more time that passed, the more objective reality succumbed to the "reality" constructed by the lonely individual's negative social cognition.

Despite their display of social skills in the study that specifically asked them to take on a supportive role, lonely students have been shown to be less responsive to their classmates during class discussions, and to provide less appropriate and less effective feedback than nonlonely students.[20] All in all, the cognitive and behavioral distortions induced by loneliness can cause plenty of trouble, and that is before the external environment gets into the act. The world is not always benign, and even at the level of the chromosome, new research shows, it can be a jungle out there. The agenda of any given gene does not necessarily align with the plans and wishes of the individual it is passing through. Many genes, fragments of chromosomes, and stretches of noncoding DNA act in their own interest at the direct expense of other genetic elements. Some damage other chromosomes in order to get themselves replicated as part of the repair process. Others disable the transmission of all other chromosomes from father to offspring, making sure that affected males pass along only the renegade element.[21]

With such intense competition as part of the human condition at every level from noncoding DNA to the North American Free Trade Agreement, what kind of disadvantages do we suffer when a sense of isolation impairs our best thinking and behavior?

CHAPTER ELEVEN

conflicts in nature

Some degree of confusion is inherent in almost all social situations, and as I've noted, some of this is internal, a function of interference stemming from the way our brains are constructed. But in the external world, love and kinship mingle with resentment and competition at every juncture.

Parents love all their children, in theory, and, at least in theory, they love each child equally. Adults in the family try to encourage brothers and sisters to show affection for each other, and to share resources equitably among themselves. But parents and children have only a fifty percent overlap in their genetic interest. Each child has his own evolutionary agenda, which begins with extracting all the parental love and resources he can. Mothers and infants squabble about the amount of time spent breastfeeding. Littermates fight over access to the nursing mother, and the runts get shoved to the end of the line. Young chicks sharing a nest will often fight and may push their weaker siblings out. In some species nestlings peck one another to death while the parent placidly observes.

In such a world, we need to be able to discriminate between genuine affection and manipulation, and to look out for ourselves accordingly. Unfortunately, owing to the kinds of impairment in self-regulation and errors in social cognition I've described, we become less well suited to these tasks when loneliness takes hold.

Robert Trivers, one of the founders of evolutionary psychology, makes a distinction between teaching, which is for the benefit of the child, and molding, which is for the benefit of the parent.[1] A quick glance at almost any playground or department store fitting room will show some parents badgering and bullying their kids with overbearing instruction. At home, a mother might try to "mold" baby Susan into taking a nap every day at three so Mom can get a break, or a father might try to "mold" Charlie into becoming an ophthalmologist like himself rather than a saxophone player like no-good Uncle Ralph.

Lonely children may be less capable than others of standing up to that kind of pressure in defense of their own interests. When we are lonely we are more likely to adopt the consensus opinion, we mimic others more intensely, and we are less likely to exhibit persistence. And when we're lonely, just as at every other time in life, we need love, a need that itself can be coercive, even to the point of persuading us to betray ourselves.

For the same sorts of reasons that natural selection maintains a certain amount of variation in the behavioral attributes of any given population, natural selection favors a certain behavioral flexibility in each of us. Thus we find that, even in those who generally reach first for kindness and generosity in the tool kit of social skills, there reside vestiges of dishonesty and duplicity held in reserve. An infant with the cleverness to deceive as she asserts herself will have a slight advantage over her parents as well as her peers. Mothers and fathers with the cleverness not to be deceived will gain an advantage, meaning that they will be able to allocate their resources more equitably across the entire span of their genetic investment—that is, among all their kids. With each advance on the part of one party or another in the game of deception and detection, the competition intensifies a notch, as do the complexity and sophistication of the neural equipment needed to keep playing.

Although the advantages of protective social connection recommend a certain degree of deference to parental advice and instruction, what we call maturation requires evaluating these influences and, eventually, learning to pay more attention to what *feels* right to us. Here, again, the lonely child is at a considerable disadvantage. To exercise subtle discrimination among one's own desires and goals, the pull of emotions, and the pressure of outside influences requires exemplary executive control—precisely the kind of self-regulation that the feeling of isolation impairs.

Even chimpanzee youngsters know how to deceive, throwing tantrums, then glancing over furtively to see if the mother is paying attention. And even in chimps, self-regulation and co-regulation also involve a sense of fairness, as well as acute perceptions of what is fair and what is not. If you doubt it, offer one chimp a cucumber slice and another a grape for performing the same task, then see how they react.

In 2006 we set out to test the impact of loneliness on responses to inequitable treatment. Our strategy involved a game in which the researcher designates one player as "proposer" and the other as "decider" and gives the proposer ten dollars. The proposer must split the money with the decider—along whatever lines he can get the decider to accept. If the decider rejects the proposal, neither player gets any money. The proposer has a natural incentive to try to keep as much as he can, but he also has to, in some sense, buy the decider's acceptance. Can he get away with offering a dollar, or will agreement require a fifty-fifty split? A dollar is better than nothing, but the fact that the proposer might arbitrarily receive nine dollars hardly seems fair to the decider who would get only one. Maybe seven dollars and three dollars will do the trick. It's all in the negotiation, and in the decider's perception of what constitutes an acceptable level of fairness.

In our version of the game, this same scenario unfolded twenty times, with the second player each time deciding whether to accept or reject the split being offered. But, being experimental psychologists, we set it up so that the "proposer" was one of our confederates, so that we could orchestrate the deals. The proposer and the decider were placed in separate rooms. Then, via a speaker system, the proposer—actually, to ensure experimental control, it was her recorded voice—made a series of twenty propositions, offering the other player a share that ranged from one dollar to five dollars. The proposed deals were carefully arranged so that either the first ten or the second ten were close to fifty-fifty. The other ten offers were clearly unfair, meaning no more than three dollars for the decider.

It will probably come as no surprise that most people are sensitive to whether or not another person is dealing with them fairly, and that they agree to accept more fair offers than unfair ones. They do this even when, as in our experiment, rejecting an offer leaves them with no reward but their pride and their sense of right and wrong. Lonely players generally followed this pattern, and lonely and non-lonely participants in our game accepted comparable numbers of fair offers. However, lonely players accepted more unfair offers than did nonlonely players. They went along more often when their partner treated them unfairly, even though both lonely and nonlonely players rated the offers as equally and profoundly unfair.

This willingness to endure exploitation even when we have a clear sense that the other person is treating us unfairly does not bode well for our chances of achieving satisfying social connections in the long run, and it can place lonely individuals at greater risk of being scammed, or at least disappointed. Over time, the bad experiences that follow can contribute to the lonely person's impression that, when you come right down to it, betrayal or rejection is lurking around every corner—a perception that plays into fear, hostility, learned helplessness, and passive coping.

Even among apes, fairness and reciprocation means trading favors back and forth. In close relationships they swap seemingly without any accounting or particular acknowledgment, but in more distant relationships they reward a lengthy grooming session or some other good deed tit for tat. And even apes have the brainpower to keep a running score. The primatologist Frans de Waal describes a chimpanzee named Georgia who was known to be unusually stingy and therefore was unpopular. Whenever there was meat to be shared, her bad reputation meant that she had to beg and solicit for a much longer time than any of the others.[2]

In humans even more than in chimps, natural selection favored genes that generate feelings of gratitude calibrated to reflect the magnitude of the altruistic action or the favor bestowed. Allowing a stranger to borrow your cell phone in an emergency is hardly the same thing as donating your kidney, or giving blood, or even sending a hundred dollars to National Public Radio. Natural selection, as well as modern commercial culture, favors those who know the difference, and who can discriminate appropriately in their social exchanges. But natural selection never gives out a blank check.

The Role of Sanctions

Researchers at the University of Erfurt in Germany and at the London School of Economics conducted an experiment that separated eighty-four participants into two different investment clubs. The experiment consisted of thirty repetitions of a three-stage process: a "choose your institution" phase, a "make a contribution" phase, and a "sanctioning" phase. Each participant in the experiment started out with twenty "units" of money. After choosing which institution to trust with their investment, these players next chose what to contribute to the collective fund, a sum that could range from nothing to all twenty units. Whatever a participant chose not to contribute went into her own private account. Whatever went into the collective pot would increase in value, but at the end of play it would be divided equally among all the members of that institution, regardless of the level of their individual contributions. Those who contributed the least, those guilty of what psychologists call free riding or social loafing,[3] would receive just as much as someone who gave her all. To keep things interesting, at the end of the contribution phase of each round all the players were told how much everyone else had contributed. They were also kept informed of everyone's earnings to date.[4]

It was in phase three, with the introduction of sanctions, that the two institutions diverged dramatically. In institution A, there were no sanctions. In institution B, each player had the options of penalizing free riders and rewarding the generous. Each player could assign a penalty token worth three money units to noncontributing members, but at a cost to themselves of one money unit. Conversely, they could reward especially generous contributors with a token worth one money unit, at a total cost to themselves of that one money unit.

At the beginning of the experiment, only about a third of the participants chose the institution that would enforce sanctions with financial penalties. However, by the tenth round of the game, almost ninety percent of the participants had elected to go over to B, the institution with sanctions. Moreover, the folks in B were cooperating fully. By the thirtieth round, contributions to A, the sanction-free institution, had dwindled to zero. The reason for the surge in B: An institution in which generous contributions to the commonwealth are the norm—a norm reinforced by sanctions—delivers the highest returns to its members.

True, free riders in sanction-free institution A earned the highest payoff in the first round. But after that their payoff sharply declined, leading eventually to a total collapse of their laissez-faire arrangement. After the fifth round, it was clear that high contributors in the sanctioning institution were earning more, which had a snowball effect on membership and earnings, as more people saw the benefits and wanted to join. With more members joining and contributing freely to institution B, the benefits of this positive social behavior became greater still.

But much of the credit for high earnings in the sanctioning institution belongs to the "strong reciprocators," those who punished others for riding free, even at a significant cost to themselves. In the first rounds there was no obvious financial benefit to slapping other participants with fines. But after several iterations, this enforcement of standards led to a continuing increase in efficiency, with more and more people becoming high contributors, so much so that the need for negative sanctions faded away.

We can see the same bias toward social cooperation at work in the real economy. Oligarchs or speculators may make a killing in a country where an elite can get away with anything and dissidents are kept down with an iron fist. But over the long haul, investors want their money in societies that are stable, and where the rule of law, also known as sanctioning, prevails. When the world looks especially topsy-turvy, having your money in a bank in Switzerland looks like a good bet. Benign social democracies like those in Scandinavia seem to do reasonably well year after year. Tightly knit communities like the Amish or the Mormons also do a good job of prospering by looking out for one another.

On the other hand, the more widely we extend our contextual circle, the more difficult it is to maintain an Amish level of consensus and homogeneity. As we move from the pair bond to the tribe, to the nation state, to humanity at large, greater variation leads to greater economic, political, and historical complexity. This makes moral absolutes—the sharia law, for instance, or the "no drinking, no dancing" strictures of Southern Baptists—harder to maintain as once-isolated communities become more cosmopolitan. As cultures encompass more variety in their mores, the absence of narrowly defined and rigidly enforced standards places even more of a burden on the brain's executive function, not only to guide self-regulation, but to calibrate, discern, and fine-tune appropriate responses. This is where the developmental plasticity of humans provided us with an evolutionary opportunity, and, once again, where the interference effects caused by loneliness can create additional burdens.

Just as in the investment club game, or in the Prisoner's Dilemma tournament described in Chapter Four, the route to achieving satisfaction in social connection is not blind and naïve altruism. Rather than a simple stimulus-response that aligns with fixed patterns, the most effective strategy is to have an enlarged cognitive capacity, carefully regulated through executive control, that can properly read signals from the environment, then try to determine what will yield the greater good—for others *and* for oneself—going forward.

A saintly selflessness would not necessarily be the most effective survival strategy at Her Majesty's Prison at Brixton, or among pirates off the Horn of Africa. Even in relatively benign environments, the winning combination is not quite as simple as the golden rule, but rather a "do unto others" bias toward generosity and helpfulness, with other alternatives kept readily at hand to avoid exploitation and abuse.

A Message That Not Everyone Gets

In his book *The Selfish Gene*, Richard Dawkins explored the evolution of altruism through something he called the "green beard effect." Any system of selfless cooperation could be undone by cheaters and free riders because they could reap the benefits of social cooperation without bearing any of the cost. As a result, they would be more likely to spread their purely self-interested genes throughout the population.[5]

To imagine a way around this problem, Dawkins came up with a hypothetical species in which a literal green beard beneath the chin would serve as a marker for an altruism gene. The beard, as an expression of the gene, would make members of the club easy to recognize, and would promote cooperative behavior among members—a genetic version of a secret society like Skull and Bones or the Masons.

Many species actually cooperate in this fashion, but until recently the only examples known were invertebrates—ants, slime molds, and such—in which the influence of shared genes (kin selection) tightly constrains all aspects of behavior. In 2006, however, researchers at the University of California at Santa Cruz reported on a species of lizard that operates according to exceedingly altruistic rules, without kinship as a factor. This species comes in three varieties—orange-throats, yellow-throats, and blue-throats—and the difference in coloration denotes different territorial behavior in the males. Orange-throated males are the preemptive militarists, annexing territories not their own. The yellow-throats are subversives who sneak into another male's territory and mate with the females. The blue-throats are a real life example of Dawkins's hypothetical green-beards. They form partnerships in which two males cooperate to protect their territories, one for both and both for one.

By tracking this behavior over eighteen generations, the Santa Cruz researchers found that the blue-throats do not go through life oblivious to changing circumstances, but rather they fluctuate between sacrificial altruism and more balanced mutualism. The trigger is the behavior of the bad guys—the hyperaggressive orange-throats. When these invaders are rampant, one blue-throat winds up putting in so much time and energy serving as a buffer for his partner that he fails to reproduce at all. In years in which the orange-throats are less of a threat, both blue-throats are hugely successful in the reproduction sweepstakes. What keeps the system going is the long-term advantages of mutual aid when the reciprocation continues over many annual cycles.

However, not even all the blue-throats got the genetic memo about cooperation. Because some lack the full DNA directive for altruistic behavior, they do not partner up. In the years in which the orange-throats drive sacrificial altruism, these loner blue-throats reproduce more successfully than the altruists, but still less well than the protected partners. In years in which pressure from the orange-throats is milder and the partnerships can be mutually beneficial, the loners do less well than either of the socially connected partners.[6]

Ultimately, then, cooperative "green-beard" behavior is the most successful adaptation for blue-throats, a lesson that harks back to Robert Axelrod's computer simulations with the Prisoner's Dilemma, to the investment game with sanctions, and to the equitable societies that remain stable and prosperous. However, not even all blue-throated lizards can be trusted to cooperate. Among humans the situation is trickier still, because those who cooperate don't wear a color-coded sign under their chins announcing their benevolence to the world. So once again, the advantage goes to those who can quickly and accurately spot the con artists as well as the good guys, the Mr. Wrongs as well as the potentially loyal husbands.

Among our fellow humans, we maintain a certain wariness of strangers and outgroup members, and all the more so in unusual or stressful circumstances. At the extreme of heightened alert—during warfare—combatants have traditionally worn military uniforms to keep it clear who is on which side. Rules of engagement and the Geneva Conventions make this discernment simpler, and grotesque as it may be, you die for your designated comrades and you kill your designated enemy. Not a pleasant situation to be in, but generally not hard to figure out. The anguish of combat is far worse when it takes place in the midst of embattled civilians, as we know all too well from nebulous military expeditions such as those in Vietnam and Iraq. Do I save the baby lying abandoned in the road, or is the baby actually a booby trap rigged with explosives?

Similarly, in ordinary life, the most difficult cooperation/competition signals to decode are those transmitted and received in the no-man's-land of social ambiguity, up close and personal. This is where the lonely suffer most from impairments in executive function. My co-worker is smiling and seems to be well-intentioned, but what if she is a wolf in Prada clothing? Do I really want to buy the hearing aid this nice young man is trying to sell me? Is my neighbor telling me the truth, or should I run this piece of paper past my lawyer?

With an impaired ability to discriminate, persevere, and self-regulate, the lonely, both as children and as adults, often engage in extremes. Sometimes, in an effort to belong, they allow themselves to be pushed around, as in our "proposer/decider" game, when a lonely adult feels resentment, but goes ahead and accepts unfair offers. This impairment is in play when a lonely child lets the big kids ride off with his new bicycle. At other times, fear might lead that same lonely child to almost paranoid levels of self-protection—such as not allowing anyone, ever, to play with his toys. Especially with advancing age, the lonely fall victim to unscrupulous salespeople who try to exploit their need for connection and their diminished ability to read social signals and detect manipulation. Individuals, young or old, who "try to see the good in everyone" without an appropriate level of caution are especially vulnerable to others who don't respond to, or display, the same kinds of co-regulating impulses that most humans share when we feel socially connected.

Strength in Numbers

Two thousand years ago, Julius Caesar might have survived the Ides of March had he been more astute in reading the false smiles of his erstwhile allies Brutus and Cassius. The great Roman general, however, was neither the first nor the last alpha to learn too late that allies can become assassins in a heartbeat. Whether among pygmy apes or prime ministers, a dominant player who shows any hint of weakness can slip from the top of the pyramid. Among chimps, younger males are always on the make, teaming up to plot political takeovers, looking for the opportune moment, which is why an injured alpha will put extra energy into showing off his vigor with bluster and bragging.

Nonetheless, as the experience of the loner blue-throats suggests, allies and coalitions provide too many advantages to be treated with a generalized suspicion that precludes cooperation. The key to successful social connection of any sort lies in accurately reading the social cues and in skillfully—and empathically—managing the relationships.

Predatory species such as wolves and chimps long ago evolved the ability to work together to corner and trap prey. Chimps are so conditioned for collective defense that they often fan out and go on border patrols even in captivity. In the wild, when traveling through areas in which thick vegetation reduces visibility, they rely on knowledge of each other's voices for identification and coordination. We humans similarly rely on unit cohesion. Military training tries to ensure that the individual soldiers will come to value the survival of this group of genetically unrelated comrades over the natural instinct of self-preservation. This is why societies reward this ultimate triumph of self-regulation with medals and citations praising the individual's "complete disregard for personal safety." It is each soldier's willingness to give his all for the others that provides the comfort, safety, and strategic advantage associated with the group.

Teamwork über Alles

Robert Axelrod's Prisoner's Dilemma tournament explored the evolution of social cooperation in terms of individuals acting alone. In 2004 the computer scientist Graham Kendall came up with the idea for a twentieth anniversary rematch that more closely resembled actual societies. This time, Tit for Tat's simple bias in favor of loyalty was to be challenged by a new wrinkle.

For Kendall's updated tournament, teams could submit multiple strategies by entering multiple computer programs as contestants. Ultimately, 223 entries signed up, with each program set to face the others in a round robin. In the original game, the two "prisoners" could not communicate about their intentions or anything else. This time the cyber-convicts could share information, performing a high-tech version of tapping out code on the prison heating pipes. As in real life, of course, even with communication allowed, there was no certainty that players were telling the truth.

The key innovation was made by the computer wizards from England's Southampton University. They entered sixty programs, each a slight variation on an overall strategy that allowed players to recognize one another and to act in concert. Each Southampton entry was programmed to execute a series of five to ten moves by which two Southampton programs could identify each other. If a program detected that another player was not from Southampton, it would immediately act as a spoiler and betray this other "prisoner." But whenever two Southampton players recognized each other, each one immediately assumed either a master or a slave role, with one team member sacrificing itself so that the other could win. The result was that while many Southampton players bit the dust, collectively the players from Southampton took the tournament hands down, placing first, second, and third.[7]

Teamwork requires accurate and timely information even when the contest is in the intimate domain of interpersonal relationship. Of course, the premium placed on information can encourage even more devious forms of deception and betrayal. To advance their own genetic agenda, newly dominant males in some species will kill youngsters not their own. In response, a pregnant female learns to solicit and have sex with the new alpha as soon as he takes office, tricking him into thinking that the child she already carries might be his. His reproductive objectives demand that he keep track of all females in the troop, any of whom may sneak away at any time to have sex with a male she prefers over the alpha. Among the !Kung we saw that violations of the pair bond for sexual fidelity were far from unusual within the human environment of evolutionary adaptation. In today's post-industrial societies, DNA testing to sort out uncertain paternity is a thriving business.

To stay on top, the alpha not only needs accurate information; he needs to refine his own manipulative use of it as his physical powers decline with age. On the other side of the equation, an upstart chimp, hoping to rise in status, learns to manipulate by dissembling. After being bested in a contest, the cagiest will engage in melodramatic displays of (sometimes faked) injury to gain sympathy and political support. Especially among humans, manipulation can include extreme cruelty, as when a tyrant showers his inner circle with largesse, then inflicts random acts of terror and brutality on those same courtiers to sustain and enlarge the illusion of indomitable, even godlike power. Those not quite up to speed in deciphering the complexities of their social universe, whether it is a Renaissance court or a corporate office, whether in the private or the political realm, will surely suffer.

The depth and breadth of this arms race of deception and detection helps explain our insatiable appetite for information about one another, the kind of strategic intelligence commonly known as gossip.[8] As the stories from *Nisa* suggest, dishing the dirt appears to have been a staple of human life since the time when we all lived more or less like the !Kung. Today, with *People*, *Us*, *Entertainment Tonight*, and the tabloid press, gossip, also known as social information, is a mainstay of contemporary media. Because information can be shared without being surrendered, passing it on costs the giver nothing, which makes it an almost inexhaustible medium of social exchange.

Enforcement

In addition to the impulse to know what others are doing, we all have the impulse in varying degrees to impose our will on what others are allowed to do. This means that negative social emotions—another staple of the tabloids—if not murder and mayhem, will break out often enough. This is all the more reason why letting the need to feel connected prompt naïve behavior is a bad idea. It is also why natural selection provided spontaneous, co-regulating behaviors to keep the peace, to promote the common good, and to hold the power of any one individual in check. But again, feeling like an outsider can leave any of us with an intense, unmet need to feel connected, which can disrupt the sensors that underlie this process.

The co-regulating, tit-for-tat logic of reciprocal cooperation does not always expect to be repaid on a one-to-one basis. Nonetheless, if you are willing to make a sacrifice for the good of the group—taking the lead on a dangerous hunt, for instance, or caring for a child while her mother finishes a task—you expect others to be willing, should the need arise, to make a contribution of comparable cost and significance. What results from this dense web of reciprocal expectation is a sense of one for all and all for one, which can lead from Paleolithic "three musketeers" to feelings of intense allegiance to an ethnic group, a religion, or a nation state.[9]

Anyone who has ever been in the military or played a team sport knows that punishing the whole group for the screw-ups of one individual is the best way to apply pressure to get the slacker to improve. The underlying adhesive of unit cohesion—the desire of each member to avoid bringing harm to his fellows, or shame and disapprobation to himself—works in the classroom, the office, and the barracks as well as on the battlefield. But the imposition of standards and sanctions is not exclusively the job of the leader. Bettina Rockenbach, the senior author of the paper on investment clubs and sanctions, told the *New York Times*: "The bottom line . . . is that when you have people with shared standards, and some who have the moral courage to sanction others, informally, then this kind of society manages very successfully."[10]

In the investment club study, punishing free riders ultimately led to higher returns for everyone. At the beginning, however, before the increased financial gain became evident, the sanctioning behavior of the "high enforcers" could be considered "altruistic punishment," meaning that members of the group sought to punish miscreants for no other reason than to promote more socially responsible behavior in the future. Altruistic punishment carries no immediate benefit to the enforcer, and it may even impose considerable cost.[11] Yet there are psychic rewards.

Brain scans suggest that the anticipation of a pleasant emotional response is what motivates such policing and social control. Using such imaging technology, researchers found activation of the caudate nucleus—another of the brain's reward centers—in direct proportion to the degree of punishment imposed on others.[12] We all know that doing good ultimately feels good; punishing other people for violations of an implicit social contract has its pleasures as well.

In the same vein, natural selection also favored conspicuous moral indignation and retribution whenever we sense that we are being treated unfairly. Emerging from the same impulse that makes a monkey pelt you with your proffered cucumber slice if you reward another monkey with a grape, the foundation of our legal system is embedded in our genes and in our cultures. So is the concept of honor, which, especially in traditional societies, can lead to duels, revenge killings, and many other forms of mayhem. These behaviors emerge from the fact that, whether driven by loneliness or by other factors, it is usually maladaptive to allow yourself to be taken advantage of. As chimps know very well, stable societies need what Robert Trivers called a "strong show of aggression when the cheating tendency is discovered."[13]

Moreover, indignation is a more potent weapon when it is put on public display. Ask any mafia don or gang leader—the more dramatic and public your show of outrage, the less likely you are to be "shown disrespect" in the future. You may be killed the following week, or you may spend the rest of your life in prison, but, so this primitive thinking goes, you will not be "dissed." In a violent culture such as the mob, minor infractions are repaid, perhaps, by cutting off a thumb; serious betrayal leads to being "whacked." The psychologists Margo Wilson and Martin Daly report that in human societies in general, when males kill males they know, they most often do so in front of an audience.[14] What this rash behavior lacks in terms of plausible deniability, it makes up for in terms of evolutionary psychology.

The same logic of social disapproval also extends to the verbal airing of grievances. Publicly accusing someone of having wronged you is an implied exhortation to your fellows to withhold their own cooperation and altruism from the offending person. You hope your outrage will lead to the offender's ostracism, with involuntary social isolation serving to enforce better behavior. But the plot thickened long ago, with protestations of righteous indignation becoming a natural ploy of cheaters.

For many reasons, then, the most adaptive strategy is to maintain both the ability to detect cheating or betrayal and the ability to carefully modulate one's response. The dysregulation caused by loneliness consigns us to the extremes of either suffering passively (responding too little) or being "difficult" (responding too intensely). Suffering in silence is no good, but neither is screaming at your office rival across the conference table. The distinctive human adaptation is to be socially cooperative in a way that allows us to optimize the advantages of the group while retaining our own individuality. This, no easy task, is the challenge we turn to in the final portion of the book.

PART THREE

---

finding meaning in connection

I remember the year eye contact stopped. It was not some big demographic shift. People just seemed to give up on relating to each other. Now this town is one of the loneliest places on earth. People are vaguely paranoid, oversensitive and self-involved. Incomes are high, the cost of living is astronomical, but everybody is in debt, living in million-dollar homes and eating take-out pizza. And when the divorce comes, the guy moves out of the house to live on his boat.

—Email from a man in California

CHAPTER TWELVE

three adaptations

A great deal of what it means to be human, perhaps a great deal more than philosophy, religion, or even science realized until very recently, is to be social. But we are by no means the only species that is "obligatorily gregarious." Modern humans evolved from a genealogical line of hominid, or humanlike, apes, a family tree that has had many offshoots. Ours branched off from the main trunk five to seven million years ago, at a time when climate change reduced the amount of dense rain forest in Africa and gave rise to a new habitat—the grassland or savannah. The fossil record tells us that more than a dozen other species of bipedal apes also emerged at more or less the same time. These distant cousins migrated with us to the grasslands, but it appears that their adaptations were less successful than ours, evidenced by the fact that they are no longer with us.[1] We continued to evolve and adapt, and they reached the graveyard of extinction.

Meanwhile, two slightly more ancient species survived by remaining in the forests—the chimpanzees (*Pan troglodytes)*, and the bonobos (*Pan paniscus*)—of which the chimps are far better known to us. Bonobos, in fact, were not discovered by Western science until the 1920s and were recognized as a species distinct from chimps only in the 1930s. Longer and more lithe than their cousins, with a smaller head and a flatter, more open face, they look a great deal like artists' renderings of the ancient proto-humans known as australopithecines, the group whose most famous fossil goes by the name of Lucy.

Even though the DNA of humans, chimps, and bonobos is more than ninety-eight percent the same, we did not "descend" from either of these forest dwellers. Instead, each species, including ours, is a variation off the main line, each with a different set of adaptations to social living, each of which has been successful in its own way. There is the chimp adaptation, the bonobo adaptation, and what I call the Third Adaptation—the human approach. Each of the three ways of managing a highly social existence has worked well enough to bring its species into the twenty-first century, but one has advanced the primate line considerably further than the other two.

By comparison with all other hominid apes, *Homo sapiens sapiens*, "the wise ones," also known as human beings, are, as a species, hyperempathic and hypercooperative. Which is, of course, not to say that everything is sweetness and light among us. Nonetheless, while our cousins stayed in the forest, we managed to colonize every habitat on the planet, then reach into outer space. Along the way, we have compiled forty thousand years' worth of cultural artifacts, ranging from cave drawings to monoclonal antibodies. Much of this progress we owe to expanded cognitive ability, more intense pair bonds, and a more intense and sophisticated level of parental investment in the young. Physically, we had the "upper hand" of opposable thumbs (better for tool use), upright posture (better for carrying things), longer legs (better for traveling long distances faster), and a shoulder better adapted for throwing. These physical attributes, co-evolving along with improved cognitive abilities, allowed us to open up vast new territories and to make the most of new resources. With tools to extend our grasp, with throwing to extend our reach, and with upright posture to lift our gaze, we developed a significantly wider field of vision. With the loss of body hair, we improved our ability to dissipate heat, which allowed our bipedal locomotion to evolve into running, and then into running for long periods of time. Combined with ever more sophisticated mental capacities—the ability to maintain the image of a prey animal when it is no longer in sight, the ability to continue to focus persistently on a certain goal for days or even years—running allowed us to move from scavenging on the savannahs to becoming competent hunters.[2]

With the expansion of our brain and our field of vision came an even wider expansion—not just of our range of habitation, but of our range in terms of the global and temporal nature of our concerns. It is this expansion that lies at the heart of the Third Adaptation. We became creatures not just of the moment, but of the future and the past. We could internalize lessons from experience, learn from our mistakes, and also plan ahead. We could defer gratification and we could keep mental accounts of treachery and of kindness extending back for generations, even centuries. With highly sophisticated and fully functional executive control, we could much more precisely sort out what served our own interests, while also taking into consideration our membership in various wider communities of interest, extending all over the world and into the future our great-grandchildren will inhabit. And thus, despite all the other human advantages, our most singularly beneficial adaptation remains the self-regulation and nuanced social cognition provided by our neocortex. The cornerstone of the Third Adaptation is executive function, without which our intelligence and physical abilities would have left us still as mercurial, unfocused, and isolated as Phineas Gage, the railroad worker who had a steel rod blown through his brain.

For all their cognitive development, perceptiveness, and expressiveness, both chimpanzees and bonobos are largely creatures of the moment. Not yet masters of upright bipedal locomotion, they are adapted to a downward gaze and limited concerns that match their limited field of vision. Although they can exhibit caring and even altruistic behavior, these social attributes, combined with their intelligence, have not moved them beyond a subsistence existence in a few isolated pockets in Africa, where they are threatened by human encroachment.

Perhaps because of this threat, bonobos live so deeply in the forest that we know very little about their behavior in the wild. (Happily, the Republic of Congo recently set aside the Sankuru Nature Reserve for them, with an area the size of Massachusetts.) Most of what we know about their social structure is drawn from observation of individuals held captive in zoos and other research institutions, and it is on the basis of these reports that bonobos have gained their reputation as the "hippie chimps," the original exponents of peace and free love.[3]

Some primatologists are still trying to open up bonobo research the way Jane Goodall did with chimp research, and in the meantime they are withholding judgment on the peace and love business. Still, the generally accepted perception is that chimps and bonobos offer a "Mars and Venus" contrast in how to manage group living. When it comes to self-regulation and co-regulation, chimps are the Marine Corps and bonobos are the natural foods co-op.

If bonobo society in the wild conforms to the social structure we see in captivity, it may be that unbridled peace and love are not quite as effective as the chimps' more muscular and competitive approach. But even if aggressiveness and competition bring the chimps more concentrated protein, those qualities have not done much for their cultural development. While humans have progressed to genetic engineering, *King Lear*, and the Brandenburg Concerto, our ape cousins are still sitting out in the rain, poking sticks into holes to catch termites.

Hypercooperative Homo Sapiens

Within the social structure of any hominid apes, whether in East Africa or on the Upper East Side of Manhattan, each individual must to some extent fulfill the needs of others, to some extent fulfill his or her own needs, and to some extent keep others' need fulfillment in check. The simple lesson that emerges from studying hominids is that the more extensive the reciprocal altruism born of social connection—the adaptation in which humans truly excel—the greater the advance toward health, wealth, and happiness.

Martin Nowak has itemized five distinct dimensions of social cooperation, each with an appropriate thumbnail description:

- Kin selection: "I will jump into the river to save two brothers or eight cousins."

- Direct reciprocity: "I will scratch your back if you scratch my back."

- Indirect reciprocity: "I will help others in order to gain a good reputation, which will be rewarded by others."

- Network reciprocity: "I will help others in order to avoid exclusion from a cooperative network in which members help each other."

- Group selection: A group of cooperators may be more successful than a group of defectors.[4]

Each of these rules applies to chimps and bonobos as well as to humans; the difference between our infinitely more successful Third Adaptation and the two runners-up is a matter of degree. For instance, even when any given Option B will benefit the larger social group, chimps can go either way, choosing Option A about half the time. By contrast, human children will almost always help others complete a simple task, spontaneously and without reward, by the age of fifteen months.[5]

In day-to-day life as well as in laboratory studies, individual motivation is influenced by the evolutionary goals of the selfish gene. If to rescue a baby chimp not his own, an adult chimp risks his life diving into the moat that surrounds his enclosure at the zoo, he doesn't have the time—or presumably the computational power—to do an elaborate calculation of the forces of kin selection, direct, indirect, or network reciprocity, or group selection. Instead, his action is prompted by predispositions laid down in his genetic blueprint.

Some researchers say that chimps have become "stuck" in competitiveness because of a lack of "social tolerance," meaning their choosiness about which other chimps they will help. Chimps know how to pull together—even literally—when there is a clear reward. This was demonstrated in an experiment in which two chimps could reach food, but only if they collaborated, each pulling on a separate rope.[6] But they were still very particular in deciding with whom they would cooperate. Social signaling may be another factor. Humans have reached a unique level of skill in learning to read intentions and cooperate as an integrated team. Dogs, after thousands of years of selective breeding by and partnership with our species, readily and easily attend to human gestures in order to find food. Chimps, although more intelligent than dogs, can only rarely do this.[7]

Which is far from saying that chimps are unexpressive brutes. Their lovable, playful side is just as real as their aggression, and both aspects of their temperament are very physical. After a kill in the wild, or even when caretakers in a zoo bring buckets of food, they collect like fans after a winning game, hugging each other, kissing, thumping each other on the back, jumping up and down. All that touching, of course, is a means of social regulation, as is the food sharing that follows. The physical celebration—think back to oxytocin—lessens tensions and promotes a cooperative atmosphere. And for the male hunters who bring back meat in the wild, distributing this high-protein luxury item is also a bargaining tool for sex.

The fact that chimps, for all their camaraderie, never reached our level of subtlety in social signaling, or our level of skill in performing cooperative tasks, was a good thing for early humans, because our competitive advantage was never based on physical strength. The average adult male chimp is five times stronger than the average man. With thick necks, broad shoulders, and large, razor-sharp canine teeth, male chimps have ferocious tempers. They also bluff by charging about with their hair on end to make them look larger than they are.

Bonobos don't go in for the macho stuff. But they too are very physical in their way of promoting a positive social atmosphere.

They too share food, and they too "celebrate" before eating it, but they don't just hug and kiss. Bonobos hoot loudly, and then have sex.

Hippies in the 1960s and 1970s were known to use sex as a way of ridding the communal atmosphere of "bad vibes," but for bonobos, at least in captivity, sex is the day-to-day currency of social affiliation, the bonobo equivalent of a handshake, a hug, or a wave goodbye. Upon greeting each other, one of the first things female bonobos do, virtually anywhere, under virtually any circumstances, is what primatologists call genito-genital (or GG) rubbing. Males and females, males and males, likewise have sex anywhere, anytime, with the young often jumping on top of them to take a better look.

Social Contracts

Whether by physical intimidation, by sexual gratification, or by more sophisticated social specialization, each of the three adaptations must achieve a level of group regulation that will promote cooperation, punish lack of cooperation, and govern the sharing of resources. Chimps and bonobos use nonverbal levers to govern fertility and parental investment, as well as submission and dominance, while also maintaining what politicians might call "feedback from the grass roots." This is how our ancestors acquired the "carrot and stick" of most interest to us, the physiological sensations of pleasure and pain that we call social connection and loneliness.

Male bonobos, like male chimpanzees, have sharp canine teeth, and they weigh on average fifteen percent more than females. Nonetheless, within bonobo society, at least according to the prevailing view, it is the females who eat first and the females who regulate food distribution. Seemingly blissed out by all their erotic activity, the male bonobos have no need to struggle for access to females, which means that they also have no incentive to compete for dominance. Bonobos freely copulate with members of neighboring groups as well, which takes away much of the incentive for male territoriality and the violence that results. Why should a male bonobo risk life and limb in raids to procure females from other groups when this sexual resource is, generally speaking, freely available? And like intermarriage among medieval royalty, the interrelatedness of troops created by this promiscuity down-regulates intergroup hostility even further.

Chimpanzee society is organized around male hunting parties and around warfare against the males from other troops. The bonobos' "commune" is relatively tranquil and mostly run by the females. And just as in human societies, each approach to regulation and social cohesion carries trade-offs in terms of costs and benefits.

For chimpanzees, the males' need to compete heavily for sex (combined with the preferences females exercise in their choice of sexual partners) has led to the evolution of males that are big, strong, and frequently brutal. The indirect result of these two social factors—male competition and female choice—is male dominance, which then feeds back into the interplay of sexual selection and natural selection. Dominant males have more and better reproductive options, and therefore it is in a female's genetic interest to mate with the biggest and the strongest, if only to increase the odds that her male offspring also will be big and strong, with the wider reproductive options that accrue to big strong males, and on and on from one generation to the next.

Unlike investment clubs, nature does not provide regular earnings reports to indicate which social strategies are most beneficial. The ultimate scorekeeper is the rate of survival of offspring. The rules are not always explicit, and the sanctions and rewards are sometimes almost imperceptible. Yet all social systems in the wild must achieve self-sustaining self-regulation, and they must do it through individual choices and co-regulation.

Within the species *Homo sapiens*, however, our intelligence and our wider perspective have opened up new areas for sanctions and rewards, as well as for another form of natural selection. Which is another reason why dominance in the form of "might means right" is inadequate for human advancement. Diversity, competition, choice, and survival of the fittest can apply to cultural values as well as to physical characteristics or behaviors. Richard Dawkins coined the term "meme" as a cultural corollary for "gene," which is to say, the unit of culture being transmitted.[8] The meme represents our human concern for purpose and meaning. Just as we humans are willing to die for our offspring or our comrades in a platoon, we are willing to die to preserve values, principles, and ideas that matter greatly to us. But by the same token, natural selection is a game in which success is measured by the greater number of units released into and surviving within the system. You cannot advance the meme "justice," then, through the meme "totalitarian behavior including torture." The propagation of memes requires adherence to the cultural values inherent within them. Which is to say: The ends cannot justify the memes.

Lonely at the Top

An often overlooked fact about dominance is the extent to which the alpha at the top of the social pyramid relies on his or her ability to self-regulate and to co-regulate. A dominant chimpanzee usually gains his top-dog position with more than a little help from his friends, cousins, and brothers. They, in turn, gain greater access to sexual privileges as a form of political patronage. So while attaining and maintaining alpha status definitely requires genetic brawn, it also depends on a genetic endowment for the kinds of executive-control functions that, as we have seen, are challenged by feelings of social exclusion: attention focus, self-restraint, impulse control, social awareness, even social sensitivity.

Alpha status depends on male-male cooperation, so even among apes, senior management requires insight, trust building, ability to detect treachery, and reciprocation. This is the only way that the leadership can establish and maintain "minimally winning coalitions" that preserve important roles, and attractive benefits, for all members of the team. As every ape learns sooner or later, a social system built on "winner take all" is never viable for long.

For chimps, regulating social interactions through constant battle comes at a price. The single male that winds up atop the hierarchy does not suddenly become immune to the competition that drives the selection process. It may be "lonely at the top," but still he has to maintain executive control. He has to be focused and attentive every moment—strutting his stuff, watching his back, playing two-against-one to maintain the alliances that will allow him to avoid being toppled by rivals.

Every claimant to the throne needs the same social aptitudes, because every up-and-comer must devote similar amounts of time and energy to managing alliances, treachery, and political triangulation. As a result—and here we get physical again—life for male chimpanzees is incredibly stressful. Researchers often see young males trembling in fear, screaming, and suffering diarrhea because of their anxiety. Not surprisingly, in chimp society, even though male and female infants are born in equal numbers, there are usually twice as many adult females as males.[9] Bonobo males are thought to endure far less social stress, and in bonobo societies the numbers of adult males and females are roughly equivalent. But again, surrounded by a tough, competitive world outside the troop, as their threatened survival shows, it may be that bonobos have taken the "easy breezy" approach to social regulation about as far as it can go.

Among chimpanzees, and not surprisingly given the price they pay and the issues at stake, hyperaggressive alphas are very clear about punishing anyone who does not show appropriate respect for the perquisites of rank. After displays of dominance, which can include rolling boulders and brandishing clubs, an alpha will often sit down and wait for his court to assemble around him. As a show of deference, underlings perform a bow that primatologists call "bobbing," but which just as easily could be called groveling, accompanied by panted grunts. If you have ever seen politicians rush over to congratulate the winner of a primary election, you know the routine. It is all part of the social contract: Dominance protocols (a form of co-regulation as well as self-regulation) maintain order, while submission and its rituals of deference (a form of self-regulation as well as co-regulation) promote harmony. The second, often neglected half of this balancing act—deference—also serves to prevent pointless battles, unnecessary injuries, and wasted energy, whether in the form of dollars in a political campaign or calories in the rain forest.

A carefully regulated balance between dominance and deference, competition and compliance keeps those at the top on their toes, while also providing something for everyone. The group achieves the kind of social homeostasis that can keep it from devolving into chaos and then breaking apart. And this complex social equation found at least temporary balance within all three social adaptations without the benefit of a UN charter, a Magna Carta, or political consultants. The rules, and the wisdom to make them work, are written in the genes and in the memes.

Calming

Among female chimps and bonobos, rank most often is based on personality and age, so there is little to fight over, and the hierarchy is largely undisputed. And yet the pressures of self-regulation and co-regulation remain strong. Even in captivity, when female chimps are brought into research labs for a learning experiment, one will always defer to the other as her superior. She will hold back, and will not touch the puzzle box, or the computer, or whatever else is offered, until the dominant female leads the way.[10]

Among bonobos, if a low-ranking female commits some offense against a dominant female's child, or grabs a piece of food that an older female had her eye on, or fails to surrender ground when a matriarch moves in to groom a male, the higher-ranking female may refuse to share food with or to accept grooming from her subordinate. This kind of rebuke can throw the younger animal into a tantrum right in front of the cold and rejecting elder. The affront is so stressful that it makes the subordinate physically sick, often causing her to vomit at the feet of her nemesis. It appears that apes do not enjoy social rejection any more than humans do.

Not just day-to-day interactions, but some of the co-regulating physiological prompts that control the larger social structure of bonobos, are based on sex. At the onset of puberty, females lose interest in the erotic play that is very much a part of their culture. They then leave their home troop to find another, a move that helps to minimize inbreeding. By the age at which playful promiscuity might lead to pregnancy, brothers and sisters have been separated. But the factor that then becomes most important for the young, migrant female is having the social finesse to be accepted by a new group. Bonobos, every bit as obligatorily gregarious as chimps or humans, cannot survive for long alone. So immediately upon encountering strangers, the young outsider engages in genital contact with older females, which leads to sponsorship by one of the matriarchs, which helps to reinforce and perpetuate the female social alliances that co-regulate the troop and hold it together.

Even within the chimps' male-dominated society, it is the females who pull the invisible strings that reinforce these hominid social contracts. Leadership depends to a large extent on the consent of the governed, and whenever a new, upstart male comes along—the ape equivalent of a nineteen-year-old boy with bulging muscles, his baseball cap on backward—the females band together to set him straight about just what sort of "dominance" they will tolerate. Even in an unstable and hierarchical society like the chimps', it does no good to knock off the top guy to take his place if then the entire troop rises up in rebellion. Overbearing leaders often have short tenures. Once again, for alpha and for insurgents alike, reading social cues correctly, and being sensitive to power dynamics, counts just as much as brawn and youthful vigor. Just like lonely humans trying to reengage with the social world in a more satisfying way, all social creatures need accurate and discerning social perceptions in order to prosper.

When female chimps sense social discord that does not please them, a few begin what is called their "woaow" bark. The first calls are tentative, as if just testing the water. When others join in, and especially when the alpha female lends her support, the calls increase in intensity until they form a deafening chorus. Without polling places or butterfly ballots, it appears that the governed are taking a vote on whether or not to give their consent to being ruled by the new guy. If the dissatisfaction is strong enough, it can lead to a full-scale revolt, with the troop driving off or even killing the would-be alpha. One leader of the troop long studied by Jane Goodall, an infamous alpha named Goblin who learned to puff up his status by banging on empty kerosene cans, was such a tyrant that he was almost murdered twice.[11]

An added incentive for chimpanzee females to work to lessen tension and promote group stability is that individual males are known to take out their frustration on everyone else. A chimpanzee male planning a power grab will spend as long as fifteen minutes warming up before launching a charge. His hair will be erect, and he will sway from side to side, hooting. Often he will arm himself with a heavy stick or rock. During this time, a female often will approach and pry the weapon out of his hand.

Females in all hominid species develop the social skills used in conflict mediation as a necessary part of motherhood. During weaning, the mother pushes the infant away from her breast, yet allows it to return when it screams in protest. The interval between rejection and acceptance lengthens as the infant ages, and the conflict turns into a major battle of wills. The youngster will try to subvert the mother with pants and whimpers; if all else fails, he may throw a temper tantrum. Again, the advantage will go to the child who knows how to successfully manipulate; on the maternal side, the advantage goes to those who know how to stand firm and avoid being taken in. Once again, this primal rejection can make the offspring so physically upset that he may vomit at his mother's feet.

Just as in human society, it is the older females who are often masters of the co-regulating behavior that helps group tension to dissipate. When two males persist in a dispute, an older, high-ranking female often will approach one of them, groom him for a while, then walk slowly toward the other and groom him. The first male usually follows her, not making eye contact with his opponent. If he does not follow, the female may tug him on the arm to make him come along. She then sits down close to the second combatant, whereupon the two males begin to groom her from either side. After a while she simply walks away, leaving them to groom each other. By that time, loud lip smacking will indicate that the erstwhile enemies are completely engrossed in their grooming. World leaders among *Homo sapiens* take note: Neither male had to take the initiative, and neither had to lose face, yet the group has been restored to equilibrium.[12]

Maintaining group cohesion and regulation among humans can be equally physical and equally subtle. Humans "have a sense" of the proper distance between two speakers in conversation. This varies among cultures, but within each culture there are clear norms. We maintain certain unspoken rules of eye contact. Staring is considered rude, sometimes even threatening. Glancing at certain parts of the anatomy is off limits because it is considered provocative or sexually aggressive. But we give no thought to any of this until someone violates the norm, and then the feeling of discomfort is palpable.

Whereas apes spend ten percent of their waking hours grooming one another, humans laugh at the boss's unfunny jokes. We also practice kindness by laughing at the sometimes goofy jokes of children, the very old, or the mentally challenged. We curry favor with our wealthy or socially prominent acquaintances, but, if we are truly socially attentive, we also change the subject the moment even the humblest listener shows signs of discomfort. When our flight hits an air pocket and the plane suddenly drops, we don't release the tension and promote group cohesion by sexually embracing our seat mate, but we may crack a joke. The unconscious and genetically biased objective in all these behaviors is the maintenance of group cohesion.

Third Time's the Charm

We have no means of assessing the social behavior of the dozen or so hominid species who have left no trace other than a few scattered bones. But for the line leading to *Homo sapiens*, we know that it was only by being able to take the longer view and to see the bigger picture that we were able to optimize social cooperation all the way up to the nation state. Obviously, humans have been more successful at this at certain times and places than at others, but the conspicuous failures of social harmony—descents into violence, misery, and economic stagnation—proves the rule. It is only when we do accommodate the Third Adaptation's broader social perspective, operating in counterpoint with personal ambition and the desire for personal gain, that we arrive at truly innovative solutions that transcend narrow self-interest. The best ideas are those that benefit the individual, the family, the tribe, and ultimately the species.

The technologies of the lever, the wheel, and fire have always been in the public domain. The wisdom of Herodotus and Hegel is available to everyone. Even corporate predators who spend decades pillaging and plundering often see the light and, in the end, set up huge foundations to do something useful with their wealth. Mother Teresa devoted her life to helping the poor of Calcutta, but not with an eye on the Nobel Prize. Sir Tim Berners-Lee invented the basic structure of the World Wide Web as a means of bringing humanity together, with no thought of commercial exploitation. And yet the human record of beneficial advancement continues to be marred by "winner take all" and "my way or the highway" thinking, including tribalism, intolerance, bloodshed, and cruelty.

Many factors—stupidity, ignorance, greed, insecurity, unresolved anger—can prevent us from consistently making good use of the broader, more nuanced, and more socially adept and beneficial Third Adaptation. Many of these causes of trouble are exceedingly difficult to address. But loneliness is one that we can begin to ameliorate in a rather straightforward fashion, especially once we realize that loneliness is not a life sentence but simply a call to repair social connections.

The motivational lesson implicit in loneliness is this: Whereas kind and generous behavior leads to social acceptance and the healthful feeling of connection, selfish antisocial behavior leads to physical decline and the disruptive pain of social isolation. Achieving connection after periods of deprivation is not easy, but as our physiology reminds us, connection is the normal state.

A distinctive level of social perception and social cognition, connection, and cooperation is at the core of who we are as a species. This means that we depend on one another not just for care and comfort but for survival itself. The socialization process may differ across cultures, but each child learns to read others, if not also to respect the rights and feelings of others. Socialization—tailoring the self in order to attain some degree of social acceptance—comes from choosing behaviors that improve circumstances for all. But it does not mean the adult equivalent of letting the big kids ride off with your bicycle.

Trying to ease the pain of loneliness and working to satisfy our need to belong often take precedence over other goals, leading people to renounce immediate gratification and self-interest in order to find better and broader long-term outcomes. But as we've seen, when social exclusion appears absolute and unyielding, the aversive feeling of isolation loses its power to motivate us. Instead, it seems to disturb the very foundation of the self. The experience of social isolation threatens our sense of purpose, which is one of the unifying factors in human development. It undermines the implicit bargain—self-regulation in exchange for social acceptance—on which personal identity is based, and which is one of the basic organizing principles of human society. Is it little wonder that loneliness is a risk factor for suicide.[13]

We have seen that feelings of isolation can cause declines in executive control and self-regulation that lead to impulsive and selfish behavior. The ability to respond actively and purposefully also declines, replaced by passivity, negativity, and sometimes even clinical depression. While our automatic responses and habits continue, as per Phineas Gage after his brain injury, our capacity for complex thought is impaired. Loneliness makes us less capable of screening out distracting cultural "noise" and focusing on what is truly important.

And these behavioral trends have snowball effects. In depriving us of self-regulation and executive control, loneliness assaults both our self-restraint and our persistence. It distorts cognition as well as empathy, in turn disrupting other perceptions that contribute to social regulation. These include our perceptions of the give-and-take of social synchronization, appropriately measured acts of deference and dominance, peacemaking, social sanctioning, and alliance formation.

The point is, we need these subtle abilities to facilitate not only our own "fit" within the group, but the group's fit overall, which is to say, a workable level of social harmony.

Loneliness diminishes the feeling of reward we get from interacting with other people. Instead, it pushes us toward an often off-putting response governed by parts of the brain associated with addiction. If I cannot read others accurately, I cannot pick up the nuances, and I cannot intuit my way to win-win solutions for the greater good. My obtuseness will lead to my not being seen as an agreeable partner. Because of my own responses, as well as the responses I elicit from others, I may become dissatisfied with my social interactions because I will not be getting the feelings of reward others receive. And the loss to me as an isolated individual may then take root and spread across my society.

Other research confirms what spurned lovers know—that when people feel rejected or excluded they tend to become more aggressive, more self-defeating or self-destructive, less cooperative and helpful, and less prone simply to do the hard work of thinking clearly.[14] On the societal level, we can see the same principles played out in distressing headlines every day.

Health, Wealth, and Happiness

Most of us eventually learn that genuine happiness over time does not equate with anything as simple as more gadgets, a bigger car, or a full belly. Happiness is not merely the opposite of pain, sadness, or discomfort.[15] Nor is genuine happiness simply a transient mood state.[16]

Years ago a classic study showed that, within two years, the happiness of lottery winners, as well as the happiness of accident victims left quadriplegic by their injuries, returned to approximately the same level enjoyed by the particular individuals before they experienced either their windfall or their misfortune.[17] Our research with older residents of Cook County has shown a similar stability in levels of happiness when tested over subsequent years, suggesting that it has a great deal to do with one's fundamental disposition. But happiness as a disposition cannot be reduced to personality constructs alone. Happiness for a member of the human species demands connection.

In our Cook County research, all the objective facts and subjective evaluations that our older participants provide from their daily lives flow into a data pool that we analyze just as carefully as we do their blood chemistry. That large body of research has been an attempt over many years to parse out the elements leading to a better life.[18]

The picture that emerged from our study is this: Having access to practical help from others was not related to happiness, but levels of loneliness and self-esteem were. Chronic stress had, as one might expect, a negative effect on happiness, but there was no direct association when the analysis included all the other variables that might influence happiness. Depressive symptoms and hostility likewise were not directly associated with happiness or the lack thereof when other variables were included. Poor health and chronic pain were only weakly correlated with happiness—presumably because people who endure these afflictions manage to adjust. The healthfulness of a person's lifestyle—factors such as smoking, alcohol consumption, exercise, and nutrition—had no measurable effect on happiness (although daytime fatigue did have a transient effect). Age did have an effect, though, and in ways that might surprise you: Older adults reported greater happiness than younger ones.

We then conducted longitudinal analyses to determine which factors predicted changes in a person's happiness over a three-year period.[19] Ultimately, we found three that filled the bill:

1. *Social connections.* Being less lonely bodes well for happiness, and happiness bodes well for one becoming less lonely over time.
2. *Household income.* Our cross-sectional data showed that household income *is* associated with happiness. However, higher household income did not predict subsequent *increases* in happiness. There is, in other words, a limit beyond which more income ceases to make you happier. We did find an association between income

and happiness, but in the reverse order: Higher levels of happiness predicted subsequent increases in income. And happiness predicted higher income at least in part through lowering feelings of loneliness.

3. *Age.* Despite common perceptions about the "miseries of old age," our data and those of the psychologist Laura Carstensen tell us that people actually get happier as they grow older. Two factors may explain this. The first is that the amygdala, the brain structure that governs our emotional responses—especially negative responses—may become slightly less reactive to negative stimuli over time.[20] As a result, older people, on average, just don't get as worked up over all the potential threats that used to bother them. And the second, perhaps more important factor is that older people, knowing that they have less time to fritter away, begin to focus on the aspects of life—human connections—that are most emotionally satisfying.[21] (See #1 above.)

If we want to turn these data about happiness into a course of action, both for individuals and for our society, we have to accept that age is what it is. Realizing that we won't live forever may help us take certain vexations less seriously and refocus our values, but I doubt that anyone would hurry the onset of age just in order to mellow out. As for the other two factors, the specific causal relationship that emerged from our data allows us to make certain recommendations.

Relying on increased income for your direct pathway to happiness is not a great idea. Anyone who watches the news, reads, or goes to the movies has been exposed to the myriad pitfalls of a ruthless pursuit of money; our data offer demonstrable evidence that, in fact, you cannot simply buy happiness. Our longitudinal analyses showed that whereas low loneliness and increased income are both *associated with* greater happiness, increased income does not *contribute to* greater happiness, nor does it lower loneliness. Instead, it is the other way around: Greater happiness, through its positive effect on social connections, contributes to increased income. Happy people become less lonely people, and people who are less lonely tend to make more money.

How is this possible? Relying on our data, we cannot say precisely. But we do know that happier, less lonely people form good relationships, including relationships in the workplace, and it may be that these good relationships, rather than happiness itself, improve job performance, increase the likelihood of receiving good performance reviews and promotions, and provide better networking opportunities for career growth. Happiness, in conjunction with lower levels of loneliness, appears also to promote more creative decisionmaking,[22] which can lead to greater financial rewards.

So what is the proverbial bottom line? Given our distinctive human adaptation—the Third Adaptation—what is the best course of action for getting the most out of life?

Well, if we were like that lonely pilgrim struggling to the top of the long mountain trail featured in innumerable *New Yorker* cartoons, and if we were able to ask the guru sitting there, "What is the key to health, wealth, and happiness?" his answer, according to our data, would have to be something like this: "You are fundamentally a social being. The key to it all is to form strong social ties that are meaningful and satisfying, both to you and to those around you, near and far."

CHAPTER THIRTEEN

getting it right

The University of Chicago is on the city's south side, and my wife and I live on the north side. Fortunately for me, the route between my home and the campus follows the shore of Lake Michigan, so when the weather is good, I sometimes ride my bike.

I am by no means the only Chicagoan who takes advantage of warm temperatures and sunshine. Often, when I'm out there, the broad, paved path along the shoreline is crowded with strolling couples, rollerbladers, skateboarders, moms and dads with baby carriages, and hundreds of other people jogging and biking.

It's usually very pleasant to glide through this sea of humanity, where everyone appears to be in a good mood, brightened not just by the weather but by the spectacular view. But the crowd also has its dangers. Each of these people or clusters of people is moving randomly, and they move at very different speeds. Sometimes they stop short for no apparent reason. Sometimes they dart to the left or the right just as unpredictably. When I am trying to make good time—as in, I have a class to teach—I have to be very attentive to the way my anticipated path may compete with the trajectories of other people.

Some bike riders simply barrel through, as if they expect everyone else to get out of their way. Others appear intimidated by all the confusion, moving so slowly and uncertainly that they themselves become a hazard. But on a good day, simply by watching for ripples in the larger pattern, any one of us out there on a bike can tune in to people's trajectories so as to anticipate, adjust, respond, and travel at a comfortable pace. When another bike rider comes along traveling at the same pace, we spontaneously fall into resonance, sharing time at the lead and cutting our way through the wind, without ever saying a word. A quick wave signals that one of us has reached our destination, and the alliance ends as quickly and quietly as it began.

To me, that kind of effortless synergy is what a life of social well-being is all about. When we feel safe and secure within our social connections, we can move along free of biases and unwarranted expectations. Relaxed and attentive, we can be in sync with the movements of others. With no expectation that we'll be excluded, our defensive, fight-or-flight mechanisms are not on ready alert. Free from all that distraction, we are able to detect more reliably whether any budding connection is promising or an invitation to the blues. Living more calmly in the moment, we can make better choices. Which has the added benefit of helping improve our larger social environment over time.

Cause and Effect

A colleague of mine—let's call him Paul—was traveling from Washington to Boston by train when he stopped in New York, then made a side trip into the far reaches of New Jersey. When it was time to leave, the friend Paul was visiting could not take him back to the commuter rail so they called a taxi. But in short order the driver got lost, and Paul missed the train back to the city. The tricky thing was that Paul was trying to connect in New York with the Amtrak service to Boston. "So now my entire day was screwed," Paul told me. But then he related what happened next:

> When I could see we were driving in circles I didn't get all worked up about it, even though there wasn't another com-

muter train for an hour and a half, and God knows when the next Amtrak connection was heading north. I counted to ten a time or two, and then I started thinking about options—like, Isn't there a bus station nearby? Don't the buses run more often than the train? So we got back on the highway heading for the next town where there was supposed to be a bus stop. This driver must have been feeling pretty sheepish, because after a while he said, "You know . . . I could take you to Newark. You can catch that Amtrak from Newark same as you can from New York." So we struck a deal. For ten dollars more he would take me halfway across the state, directly to my real objective—the northbound train—faster and in more comfort than the original plan.

So I sat back and relaxed, and then, just by chance, we started talking baseball. It turned out he used to live in Boston and, like me, he was a lifelong Red Sox fan. He had some great stories I'd never heard before about Cy Young and the 1903 World Series, so as it turned out we had a great time. If I'd yelled at him for getting lost, I think the best I could have hoped for was to find the station we were looking for in the first place, then cool my heels for an hour and a half waiting for the next commuter train.

When ancestral humans encountered cooperative behavior, they experienced the sensations we now call affection or trust. When they confronted deceit and betrayal, they experienced the sensations we now call hostility, distrust, or anger. As I've suggested through dozens of examples, when we feel isolated we also feel embattled, which leads to less robust health, less enjoyment in life, and less of an ability to collaborate to find winning solutions. When we feel satisfied with our social connections, we feel safe. When we feel safe, we can think more creatively. We also anticipate and more often experience positive emotions, which, aside from their long-term physiological benefits, provide immediate and persistent psychological uplift. That boost in mood affects our subsequent behavior toward others, which, in turn, affects how others behave toward us—which, once again, encourages creative collaboration. Cause and effect cycle back and forth, and the positives continue to ripple outward in a widening circle.

Paul also told me how he had started his train trip in Washington the previous day. Waiting at the station, he had ordered a sandwich that turned out to be twice the size he needed. He didn't want to carry it with him, and he couldn't bring himself to throw it away. He didn't dare offer it to any of the other customers, because he assumed they'd think he was nuts. Then it occurred to him—this was a train station in a large American city:

> I began scanning the big open space, and within about a minute I spotted this scruffy-looking guy walking along, his face lined with dirt, his clothes in tatters, and I literally went running after him. When he turned around, I leaned forward with the sandwich and said, "Do you want this?" He looked up at me with this scowl, a little skeptical. But then he took it, and then he kind of nodded, and that slight look of recognition from him—it was like we connected for just that one instant. I knew full well he was the one doing *me* a favor. But if I were religious, I'd say this small exchange left me feeling blessed. And I swear it was that sense of blessing that carried over the next day in my reaction to the cabbie who got lost.

Who can say if Paul's assessment is correct? But I do know this: Some of the most serene and joyful people in the world take care of others all day, every day. Among these are hospice nurses who work to provide AIDS patients with a dignified and pain-free death. These women and men are surrounded by sadness and pain, and yet the connection they make with their patients is about as real as connection can get. The good feeling of meaning, closeness, and caring that these nurses exude—a sense of "blessing"—is also called the "helper's high." But it is not restricted to those in the helping professions. The same positive emotions can result from what the bumper stickers call "random acts of kindness" at any time during a busy day.

Feeding Back

Through social cognition, self-regulation, and co-regulation, each of us contributes to the social reality that produces the sensations that others reflect back to us. Over time, these sensations either enhance our health or wear it down, but they also help create our immediate social enviornment. By increasing the frequency of certain memes or cultural values, they help create our large-scale social reality as well.

Minimizing negative emotions can enhance the positive trend in such feedback loops, but even small efforts to introduce the positives can take us further faster. Some people, when they arrive at a tollbooth on the expressway, pay the fifty cents or the dollar for the driver in line behind them—a random act of kindness. I know a woman who, when she's having a bad day, leaves change in the soda machine for the next person to find. This may seem trivial, but studies show that the beneficiaries of such small gestures are indeed more likely to help someone else shortly thereafter.[1] Knowledge of that fact—the woman I'm describing, truth be told, is a research psychologist—helps boost my friend's own spirits. To echo Henry Melvill once again: "Our actions run as causes and return to us as results."

In the field of complex adaptive systems, scientists refer to the Butterfly Effect, whereby the wind displaced by the flutter of a butterfly's wing in Africa might initiate an immensely involved string of consequences that alter the weather over Europe days or weeks later. This particular example may be something of an exaggeration, but it isn't just a metaphor. Using supercomputers, researchers can actually work out the details that allow simple causes to interact, compound, and amplify to yield complex and profound results. In more technical terms, the Butterfly Effect is called "sensitive dependence on initial conditions," and it reflects the way that small-scale events interact with large ones. The more dramatic the small-scale cause, the more immediate and more easily measured the large-scale results.

Many families deal with grief through the supremely altruistic act of donating the organs of their lost loved one. The parents of a murdered child, more often than you might think, plead leniency for the killer. In 1993 Amy Biehl was on a Fulbright Fellowship, helping register black voters in apartheid South Africa. While driving through a township to take three of her black colleagues home, she was surrounded by a mob from the Pan Africanist Congress who dragged her from her car, hit her with a brick, and then stabbed her to death. Four young men were convicted of killing Amy and sentenced to eighteen years in prison. When they applied for amnesty under South Africa's Truth and Reconciliation Commission, Amy's parents spoke in their support. The Biehls also set up a foundation in Amy's name to continue her work on behalf of South Africa's oppressed majority. One of the projects supported by the Amy Biehl Foundation is a bakery where two of the men convicted of the killing were given jobs.

How can bereaved parents avoid hatred and behave so altruistically? The old adage says that revenge is sweet. Another says that, the necessity of sanctions notwithstanding, to forgive is divine. In terms of personal health as well as of social consequences, the evidence shows that we gain when we make forgiveness not only divine, but human.

Under extraordinary circumstances, human beings are willing to make great sacrifices in service to the interdependence that gives meaning to their lives. Field Marshall Sir William Slim, commander of British forces in Burma during World War II, described the dominant feeling on the battlefield as a sense of loneliness. He also said that the only way to victory is through morale, and that the basis of morale is the individual soldier's refusal to betray his comrades.[2] Like the soldiers he describes, each of us combats our loneliness by committing ourselves to others.

Siegfried Sassoon, one of the many literary figures who served as a British officer in World War I, expressed the loneliness of battle as well as the sense of connection in more fatherly terms:

> I looked at my companions, rolled in their blankets, their faces turned to the earth or hidden by the folds. I thought of the

> doom that was always near them now, and how I might see them lying dead, with all their jollity silenced, and their talk, which had made me impatient, ended for ever . . . my own despondency and discontent released me. I couldn't save them, but at least I could share the dangers and discomforts they endured.[3]

If you read the book or saw the film *Black Hawk Down*, you probably remember the actions of Master Sergeant Gary Gordon and Sergeant First Class Randall Shughart, the U.S. Army Rangers who insisted on being dropped into the streets of Mogadishu to protect a downed helicopter crew, men who, themselves, had gone in to try to save ground forces who had been pinned down earlier. Gordon and Shughart knew full well that hundreds if not thousands of armed rebels were descending on the crash site. In fact, they knew that what they were volunteering for was, essentially, a suicide mission, but they did it anyway. They were both killed—it was their bodies being dragged through the streets of Mogadishu that outraged the world. But by their actions they managed to save the life of the downed pilot, Chief Warrant Officer Michael Durant.

Something other than a death wish drew Gordon and Shughart to such an act of self-sacrificing "social connection," and that kind of heroism is truly exceptional. That's why these two Army Rangers were posthumously awarded the Congressional Medal of Honor. And yet, when London was bombed in the summer of 2005, and when New York and Washington were attacked by terrorists in 2001, hundreds of ordinary citizens showed exemplary courage and concern for others, working side by side with the police and firefighters. There were numerous stories of rescue crews at the World Trade Center who simply turned off their radios so that they could not be ordered out of the building while survivors might still be found. Many of these rescuers died trying to save others. But there were also ordinary office workers who, instead of running for their lives, stayed behind to carry disabled colleagues down dozens of flights of stairs. Some of the most gut-wrenching images from that day were of co-workers—perhaps even strangers—who, when they knew there was no longer any hope of escaping the fire, joined hands as if to comfort each other, and leaped to a less painful death.

Then again, crisis can bring out the worst in people as well as the best. In 2005 we saw the unraveling of the self-enforcing social compact of Tit for Tat in the breakdown of order in New Orleans after Hurricane Katrina. But in the "if it bleeds it leads" world of journalism, negative stories get the most air time. Acts of charity during that crisis were so common that they could be ignored—except for the odd, extraordinarily colorful event that could be tagged onto the end of the television news. But being socially connected does not usually involve the heroism of an Army Ranger or the selflessness of a Red Cross volunteer during a flood. In day-to-day life the experience is usually far less dramatic, and it does not necessarily lead to testimonial dinners, plaques, or being knighted by the queen.

A Cure for Loneliness?

When people first learn about our research demonstrating the physiological basis of loneliness, they sometimes ask if pharmaceutical companies will ever produce a remedy in the form of a pill. When they learn more, they realize that, for most of us, there is no need for a chemical fix. Some individuals caught in the feedback loop of loneliness and negative affect, when they begin to focus on changing their social perceptions and behaviors, might benefit from medications to first bring their depression or anxiety under control. But once again, loneliness itself is not a disease; feeling lonely from time to time is like feeling hungry or thirsty from time to time. It is part of being human. The trick is to heed these signals in ways that bring long-term satisfaction.

Moreover, when it comes to remedies for loneliness, cognitive and behavioral modifications are readily available. Even on the level of body chemistry, the naturally occurring flood of hormones and neurotransmitters that produce the soothing comfort of connection—including the "helper's high"—can be made available to anyone, without a prescription. Samples of this chemical uplift are available each time we extend ourselves to others with an act of generosity, even if it is only a few coins left in a vending machine or half a sandwich given to a homeless person. Just as thirst is the prompt that reminds us to keep the body hydrated, loneliness is the prompt that reminds us how much we depend on one another. Positive psychological adjustment is immediate reinforcement and reward.

The degree of social connection that can improve our health and our happiness, as well as the daily experience of everyone who comes in contact with us, is both as simple and as difficult as being open and available to others. We sometimes refer to people who are low in loneliness as the "socially gifted," because what they have is truly a gift. But this is a gift that you and I can extend to ourselves even as we share it with others. I mentioned earlier that people high in social well-being are usually happy in marriage and high in emotional intelligence, but they are not always the leaders or the stars. Early in life they are no more attractive, intelligent, or extroverted than their lonely counterparts, and they do not spend every moment volunteering at soup kitchens or reading to the blind. The characteristic most common among those low in loneliness is a full availability to whatever genuine social interaction is appropriate to the moment. They are able to make full use of what I have called the Third Adaptation, because they are free to seek out and fully contribute to social situations and relationships. They are considerably less likely than others to let their baggage or their behavior cast a pall over any gathering. They are, in fact, more likely to elevate the general mood, but not necessarily by talking the most or by running the show. More often, they contribute through quiet encouragement of whoever is motivated to speak or lead. In an ambiguous situation, they are more likely to give another person the benefit of the doubt. Without being a doormat for exploitation or abuse—again, careful discernment is also part of the Third Adaptation—they are more likely to forgive. But these people are not a breed apart. The essential point here is that any one of "us" has the capacity to become one of "them."

Like many of those who get stuck in loneliness, some of the

socially gifted are actually quite shy. Some have a threshold for connectedness that predisposes them to feel the pain of disconnection very acutely, and for them, shipping out to manage offshore operations in Singapore might not be the best career move. On the other hand, a particular predisposition does not mean that anyone is fated to remain in the old neighborhood near Grandma and Uncle Fritz forever. With full access to the executive brain, anyone can assess his or her own comfort level, then make appropriate decisions in light of that genetic predisposition, with appropriate adjustments over time. Knowledge of the function of loneliness helps.

Accepting the Gift

Just as some individuals troubled by loneliness may benefit from medications to deal with depression or anxiety, some may benefit from seeing a clinical psychologist or psychiatrist to address the accompanying psychological issues that can reinforce their sense of isolation. The specifics of that kind of intervention are beyond the scope of our research, and, frankly, beyond the scope of my professional expertise. But once again, even *chronic* loneliness is not a "mental disorder," although it can put us at risk for depression.[4] Millions of people who feel a painful sense of social isolation do so because they have a normal human need for social connection, as well as a very normal adverse reaction to disconnection, real or perceived. Life has simply thrown them a curve, depriving them of the connection they need, creating the sense of threat that generates the negative affect—fear, anxiety, hostility—that often turns their sense of isolation into a persistent reality. But even as our emotions may roil, our thoughts are something we can learn to control. By reframing our cognitive perceptions, we can begin to change our lives.

Changing cognitive and behavioral habits does not require rooting out each and every one of the deep psychological hurts that may have shaped us over the years, but it does take practice as well as patience. The most difficult conceptual hurdle for people in the throes of loneliness is that, although they are going through something that feels like a hole in the center of their being—a hunger that needs to be fed—this "hunger" can never be satisfied by a focus on "eating." What's required is to step outside the pain of our own situation long enough to "feed" others. Which, of course, does not always involve handing out leftover sandwiches at train stations.

Reaching out beyond one's own pain sounds like a tall order. That is why the road to success begins with small steps and modest expectations.

A woman named Susan lived in Rome for a few years, working for an international organization that was quite hierarchical and unwelcoming. Although she is naturally gregarious and makes friends easily, she found herself in the unusual situation of feeling very isolated. Her colleagues made it clear that they had no intention of letting her into their circle, so she had to improvise:

> In the market there was a cobbler. He was so nice and I was so lonely . . . I remember one time taking shoes in there that didn't really need to be repaired. We did not communicate terribly well, because my Italian was so bad. But there was good intention on both sides. Perhaps he was lonely, too. He had a photograph of five young men, perhaps eighteen or so in this photograph, and they were in Naples, and they were young bucks, toward the end of the war in Europe. They were all wearing their undershirts—the sleeveless kind. I loved this photograph, and I asked him about it and he said, "Yeah, that's me there." And then later when I went in, he showed me an equivalent photograph that he'd found of the same five men, wearing those same kind of shirts, only they were now in their seventies. And when I left Italy, he gave me a copy. I was deeply touched. Because, in the end, we had probably spent only a total of twenty minutes talking to each other.

Susan did a very simple thing: She showed genuine interest in another human being, expecting nothing in return. That's all it took to make a meaningful connection, which, at least briefly, improved life for each of them. Was she intentionally "feeding" the cobbler when she dropped by and tried to relate in her faltering Italian? No, but neither was she invading his privacy to demand anything from him. She made an opening move by asking about the photograph—which gave him the opportunity to respond generously and continue the exchange, or to pull back politely and stick to business. She respected his needs and his boundaries. Her warmth and availability, along with this respect, prompted him to share something of himself, which made possible a simple, transient connection that seemed to provide comfort for them both. If there had been a context for them to establish a deeper and more enduring friendship, perhaps they could have. But as for the basics of social connection—that's the process in a nutshell.

To someone for whom loneliness has become a chronic problem, the "simple" thing Susan did may not appear simple at all. And this business about the need to feed others when we are feeling hungry ourselves is counterintuitive. "Go out and see what other people need" can leave a lonely person screaming in protest: "But I need attention! It's my turn now . . . I need payback for my miserable childhood," or "I need payback for my miserable first marriage!" Someone in the grip of loneliness may say, "I take care of everyone down at the office! I'm tired of it. I need someone in my life who'll take care of me."

"Feed me first! Take care of me!" makes a lot more sense from the narrow perspective induced by the pain of loneliness. Unfortunately, it is not a formula that works. It can be an unregulated, counterproductive response, like Sheba's pointing to the larger pile of candies. Letting go of the hope that "feed me first!" will work takes time and effort. This is when small doses of positive reinforcement, small infusions of the "helper's high," can both overcome resistance and demonstrate the promise of what can follow once we are willing to change our perspective.

Linda Fried at Columbia University established a program in which elderly residents in Baltimore are paid a small stipend to assist students in the public schools. An elderly volunteer, for instance, might give a student extra help with reading—help the teacher would like to give the student but simply doesn't have time for.

Fried found that the students benefit from the tutoring and from the attention and concern provided. But her research also shows that the elderly volunteers clearly benefit in terms of their health and well-being. Being of service in this fashion adds to the purpose, meaning, and satisfaction of their lives, and it gives them strongly positive physiological sensations in the moment. The physiological reward—also known as pleasure, sometimes known as the helper's high—can be an incentive to continue and even expand the helpful behavior. Over time, that pleasure may even compensate for, and grant us distance from, sources of lifelong emotional pain.[5]

Reinforcing Change

We are all told, in childhood, to share and to "do unto others." It sounds simplistic, like a Sunday school lesson; it doesn't sound like behavior that fits into the adult, workaday world. Certainly it does not sound like advice based on hard science. And therefore this wisdom, which should be a principle to guide us, we dismiss as a cliché. As a result, we get caught up in our problems and the confusion of our tortured perceptions and we don't practice what we know to be wise and true. The need to put simple truths into action is why various traditions came up with reminders like saying the rosary, meditating, repeating mantras and affirmations, as well as direct admonitions such as Jesus' injunction to "Feed my sheep." Real change begins with *doing*, and what may seem like silly reminders may be exactly what it takes to get us to do what needs to be done, in the moment, every day.

The study in which we manipulated lonely feelings through hypnosis shows that perceptions of social isolation are not immutable. Subjective perceptions can be reframed, which brings us back to the three structural elements of loneliness that we described in the first chapter: a genetically biased vulnerability, the need to self-regulate, and social cognition. We cannot change our genetic bias. But even when loneliness has compromised our ability to self-regulate, we can change certain aspects of our social environment, beginning with the ripples that emanate from our social cognitions.

Still, loneliness often imposes the added hurdle of learned helplessness, which leads to passive coping, and sometimes it takes a jolt to overcome that inertia.

A young father named Dave had suffered various forms of neglect as a child and was estranged from his family. In his twenties he suffered from clinical depression as well as severe loneliness, compounded by an early, unhappy marriage he entered into for reasons he says he does not fully understand to this day. Having his own child did not suddenly make those problems go away, but, as Dave puts it, "It showed me that I wasn't emotionally hopeless, or helpless."

Dave talks about getting to know his son on the night the little boy was born, sitting in a rocking chair, holding the baby, and feeling a new beginning, a new life in which it was very clear who was supposed to take care of whom. The physiological sensations of close connection were a new, powerful incentive for him to open up, commit, and give. "But the real deal," Dave told me, "was my son's first birthday party."

> We had invited some other parents with little kids, and then other friends, so it was mostly a room full of grown-ups. My son seemed to be having a good time in his one-year-old sort of way, toddling around, hanging onto the furniture. My wife stepped into the kitchen to see about something, and I was keeping an eye on him, but for a minute there I could tell that he'd lost sight of me. This look of real concern came on his face as he scanned the room filled with very large people. He was definitely getting worried, and I could see that if this anxiety went on much longer he was probably going to start to cry. But then he spotted me, and he burst into this big grin, with maybe one or two teeth showing. He put up his arms and he took a step toward me and I reached down and scooped him up and I was about ready to cry myself. It was the weirdest thing, but I swear this was the first time in my whole damn life that I ever felt fully loved and fully accepted by anyone. I mean, here is this perfect little being, this beautiful little kid, and he is in a

room surrounded by people. But he doesn't want just anybody. Of all the people standing around, he chooses me to connect with. He sees me and he's reassured. He sees me and he wants *me* to pick him up.

Sharing this moment of connection with his son did not mean that Dave was home free, but it did provide a visceral sense of what had been missing from his life. It also motivated him to get some help.

He didn't have a lot of money for long-term psychotherapy, but Dave belonged to a health maintenance organization that covered ten sessions with a counselor over ten weeks. That's not much time to probe the dark secrets of one's childhood in the classic Freudian tradition. Which is one reason that the therapists at his HMO followed the more pragmatic approach called Cognitive Behavioral Therapy.

CBT is an evidence-based, cost-effective method of redirecting emotions by modifying everyday thoughts and behaviors. It proceeds by questioning and testing assumptions or habits of thought that may be unrealistic or even damaging. It then encourages individuals to try out new ways of behaving, helping them to gradually tackle activities that they might otherwise avoid. CBT often employs techniques of relaxation and distraction, as well as having patients write down their thoughts and feelings. This journal-keeping is an opportunity to analyze when and how irrational beliefs—including, in the case of loneliness, the belief that we are and always will be socially unwanted or rejected—creep into our heads and color our perceptions. The journal often includes a three-column log in which people who want to change negative thoughts write down the activating event, the negative belief, and the consequence of that belief. For example, having a tape in your head endlessly repeating "Everybody hates me" or "I am a fat failure and a piece of garbage" is not going to help you get through your day. The first step is to recognize that indulging negative thoughts is a serious and harmful business. This kind of habitual thinking really matters because it can create self-fulfilling prophecies.

The next step is to examine that thought or belief by searching for hard evidence to back it up. The idea is to take your own internal negative statements seriously enough to examine them instead of simply repeating them. "Am I literally a piece of garbage?" When we realize that the repeated statement is not literally true—that it is often a profound distortion of reality—the only logical choice is to reframe it. What follows is to learn to identify this negativity whenever it creeps in, and then to learn to pull the plug on it the moment it appears. "Is it literally true that everybody hates me? No? Then why do I keep saying this to myself? Let's recognize the habit, and the harm that it causes, then stop it."

I doubt that anyone chooses to think in such a self-punishing way; but from time to time, any of us can slip into it. What we need to remember is the evolutionary advantage supplied by our advanced human intelligence, which is the ability to consciously redirect our thoughts. "Yes, I am not as sociable as I would like to be, but that's a far cry from 'Everybody hates me.' Some people actually like me. My mother even loves me! I should be able to change the way I relate to others."

But effective change requires more than altering our thought patterns. It requires different ways of behaving, which for the lonely, who already may be hemmed in by feelings of threat, can be an especially frightening prospect. That's why we need to ease into it with baby steps that offer the maximum in positive reinforcement at every stage.

In the next few pages, I am going to offer a few speculative suggestions based on our research findings and also inspired by an insight from the twentieth-century philosopher Reinhold Niebuhr. He tells us that "human beings are endowed by nature with both selfish and unselfish impulses," but that "man's reason endows him with a capacity for self-transcendence."[6] Self-transcendence may be ultimately what each of us is seeking as we try to connect with others. But the simple steps on that path need not sound nearly so grand.

The key is to ease your way into it, an idea that inspired an acronym—EASE—that might help you, or someone you care about, remember the four simple steps that follow.

EASE Your Way to Social Connection

E FOR EXTEND YOURSELF

The withdrawal and passivity associated with loneliness are motivated by the perception of being threatened. To be able to test other ways of behaving without that feeling of danger, you need a safe place to experiment, and you need to start small. Don't focus on trying to find the love of your life or to reinvent yourself all at once. Just slip a toe in the water. Play with the idea of trying to get small doses of the positive sensations that come from positive social interactions. The simplest moments of connection, especially when they involve "feeding others," carry an emotional uplift that does not require taking a pill, working up a sweat, or eating truckloads of cruciferous vegetables. Just don't expect too much all at once.

You may want to begin your experiment by reaching out in simple exchanges at the grocery store or at the library. Remember, if you do so, not to place any expectation on the other person. Just saying "Isn't it a beautiful day?" or "I loved that book" can bring a friendly response that makes you feel better. You sent out a small social signal, and somebody signaled back. But what if the response isn't so friendly, or you get no response at all? Maybe the person to whom you say something nice is having a rotten day. Maybe he or she is worried about a sick child, or just got an overdraft notice from the bank. A million and one factors that have absolutely nothing to do with you can influence people's moods and reactions. That's why it is important, when you begin to practice this new behavior, to make no assumptions, and to limit your objectives. You may not find that simple moment of shared human contact every time you reach out. And when you do find it, you will not necessarily have found a new bosom buddy. You need to proceed more like the birder who sees a Yellow-Eyed Junco. You feel the good feeling, mark it down on your life list and move on.

To improve your odds of eliciting a positive reaction—and to reduce your odds of being disappointed—you may want to confine your experimental outreach to the somewhat safer confines of charitable activities. Volunteer at a shelter or a hospice, teach elders how to use computers, tutor children, read to the blind, or help with a kids' sports team. You will not necessarily receive gratitude and praise for your good deeds—that's not what you're after—but it is also unlikely that you will receive scathing social punishment. There will be no big scene of fulfillment in which you are at long last voted football captain or prom queen, nor will you immediately fall into a relationship with a movie star. But you may begin to feel the positive sensations that can reinforce your desire to change, while building your confidence, while improving your ability to self-regulate. Even "small talk" about sports or the weather, when it is welcomed and shared, can be a co-regulating, calming device, and the positive change it can bring to our body chemistry can help us get beyond the fearful outlook that holds us back.

A FOR ACTION PLAN

Some people view themselves as adrift on a genetic and environmental raft over whose course they have no control. The simple realizations that we are not passive victims, that we do have some control, and that we can change our situation by changing our thoughts, expectations, and behaviors toward others can have a surprisingly empowering effect, especially on our conscious effort to self-regulate. A second inkling of control comes from recognizing that we have latitude in choosing where to invest our social energy. And as we saw in our discussion of the Butterfly Effect, it does not take an enormous change to alter one's course and destination dramatically.

Charitable activities enable us to put ourselves in the social picture with less fear of rejection or abuse, but even here some discretion is in order. Coaching kids' soccer requires at least a little knowledge of the game, but being manager or assistant coach often requires nothing more than a willingness to show up and pass around the Gatorade and the orange slices. Trying out for the community theater production could be awkward unless you really have acting or singing talent, but the theater group would probably welcome you with open arms if you volunteered to help backstage or in the ticket office. If you're shy with people but love animals, volunteer at an animal shelter. The animals will welcome you immediately. When you feel ready to reach out more to the humans around you, you can safely assume that the other volunteers share your interest in animal welfare, which gives you a natural basis for conversation, perhaps even connection.

When people feel socially connected, molehills are not mountains (grape jelly is simply grape jelly, not necessarily a sign that someone doesn't care), and most of the time, a mistake is just a mistake, not an assault on one's dignity, importance, or personhood. A less embattled way of seeing the world can help to generate a more easygoing, open nature that helps make conflicts dissipate rather than persist or escalate. The people we need to think of as models are not necessarily the most beautiful, the most fabulous, or the most socially dominant. Social connection is not a popularity contest, and the goal of change is not to win on *American Idol.* The goal is to be sufficiently secure within ourselves that we are free to genuinely focus on, and thus connect meaningfully with, others.

If you are someone who craves the warmth and connection of a very small circle, that is perfectly fine. Similarly, even when we want more social connection than we have, we may still need a little space around us, time to ourselves. That, too, is perfectly okay. We simply need to be aware of our level of need for connection and be up front about it—both with ourselves, and as we try to get to know others. Especially when you are seeking to connect with a "significant other," the challenge is to find someone who is equally comfortable with the level of togetherness that works for you.

Another aspect of developing an action plan is to remember that doing for others does not mean letting them exploit you. This is where the ability to perceive subtle distinctions provided by the brain's executive function is most vital.[7] Your action plan needs to include wariness and avoidance of those who seek to manipulate your fears and feelings. Healthy, sustainable relationships are based on willing reciprocity, not exploitation. So if your eager new friend suddenly wants to borrow money or use your car or sleep on your couch for a couple of weeks, this is probably negative data that justifies wariness, perhaps even looking elsewhere for companionship.

Feeling lonely also makes us fall victim to our own eagerness to please. Social connection does not involve superhuman strength. Committing to doing too many things for too many people in an effort to open ourselves to connection can instead make us feel overworked, stressed out, and faltering. The whole point is to be *merely* human—available to the common bond of humanity. Nor does anyone say that you have to become a long-suffering saint. Instead, the most adaptive model is an openness to engagement combined with realistic expectations, accurate perception of social cues—including cues that suggest caution—and realism about the type and number of commitments to take on. That may sound like a lot to manage, but when our executive brain is not distressed by feelings of isolation and threat, it is up to the task.

S FOR SELECTION

The solution to loneliness is not quantity but quality of relationships. Human connections have to be meaningful and satisfying for each of the people involved, and not according to some external measure. Moreover, relationships are necessarily mutual and require fairly similar levels of intimacy and intensity on both sides. Even casual chitchat—like Susan's conversation with the Italian cobbler—needs to proceed at a pace that is comfortable for everyone. Coming on too strong, oblivious to the other person's response, is the quickest way to push someone away. So part of *selection* is sensing which prospective relationships are promising, and which would be climbing the wrong tree. Loneliness makes us very attentive to social signals. The trick is to be sufficiently calm and "in the moment" to interpret those signals accurately.

I am hypoglycemic, which means I occasionally suffer falls in blood sugar that leave me ravenous. When that happens, I want and need food, no matter if it's a candy bar, a Big Mac, or a pad of butter. When I was younger, I indulged my sudden cravings by eating whatever I could get my hands on at that moment—which was usually the candy bar or the fast food. I would feel better, my blood sugar—especially after a candy bar—would spike, but then it would dive again. Eventually I learned to exercise self-control. I began to plan ahead so that lack of food did not make me feel lightheaded. Part of the planning also was to arrange to have more nutritious food that wouldn't precipitate another blood sugar crash shortly after I ate. Careful selection, guided by self-regulation, was the key lesson.

In the same fashion, we all need to learn that being drawn to someone's physical appearance or status is not a good basis for a deep connection. Compatibility and sustainability depend far more on such things as common beliefs and being at compatible stages in life. When it comes to dating and marital success, the data show that similarity ("birds of a feather flock together") trumps complementarity ("opposites attract").

Deciding how to search for birds of your own feather requires selection as well. For those who tend to be more quiet than talkative, finding someone who is also comfortable with silent companionship may be a good idea. Enthusiastic readers, especially shy readers, are more likely to find people to connect with at an author's appearance at a bookstore, or by working in a literacy program, than by going to a dance club. How you should go about trying to meet people depends on what kind of people you want to meet.

You may have seen advertisements featuring Neil Clark Warren, the founder of eHarmony, one of the most successful online matchmaking services. His system—for which (full disclosure) I am a scientific consultant—is not based on first-glance looks and sex appeal. Instead it is built around a questionnaire involving 436 separate items to determine values and interests. He developed this instrument by intensively interviewing more than five thousand married couples. After finding out what bound these people together, he built a predictive model based on 29 different dimensions of compatibility. These include values, character, intellect, sense of humor, spiritual beliefs, passion, and spontaneity.

What really works, according to Warren, is not a focus on either party's objective features, or on what either party purportedly desires, but on the actual match between the two as potential partners—the way they click as a team. Some people have criticized his system for not paying enough attention to physical chemistry, but over the long haul, feeling close sexually requires feeling close psychologically. So over time, physical attributes matter far less than psychological intimacy. The arousal you need may be induced more easily by laughing together at a Marx Brothers movie than by having a partner with a beautifully toned body.

E FOR EXPECT THE BEST

Social contentment can help us to be more consistent, generous, and resilient. It can make us more optimistic, and that "expect the best" attitude helps us project the best. According to the logic of co-regulation, then, social contentment is more likely to elicit warmth and goodwill from other people—such is the power of reciprocity. With practice, any one of us can "warm up" what we present to the world. We have more control over our thoughts and behavior patterns than we may think, but then again, no one can exercise total control of interpersonal relationships, any more than we can force an immediate and complete turnaround in the way others see us. While we wait for the change in us to register in the world around us, fear and frustration can push us back into the critical and demanding behavior associated with loneliness. This is when patiently focusing on the small physiochemical rewards of reaching out to feed others can help keep us on track.

There are risks in letting go of self-protective, isolating behaviors. People hang on to defense mechanisms because, at least for the short term, the defenses seem to do the job. But the evidence shows that the temporary "protection" provided by defensiveness comes at a high long-term cost.

The need for patience does not end once we begin to find greater happiness in our relationships. Even if any of us were perfect, inevitably the other people we come to know will have different perspectives. The prototypical wedding vows, "for better or for worse, in good times and in bad," are a public proclamation of the ever-present likelihood of interpersonal friction. Even the best friends and the partners in the best marriages will disagree and hurt each other from time to time. The secret to success in the face of this reality is not to magnify the moments of friction by overinterpreting them.

Occasionally, people who are trying to be positive and giving toward others find themselves feeling beaten down or fatigued. Interacting with a friend or a spouse who is having a hard time, who perhaps has fallen into depression, may cause you to feel depressed as well. The drop in your own energy level may be an important signal that you and your friend or your spouse may need special, professional assistance. It is also possible to go too far with feeding rather than being fed. When this happens, you need to find a way to bring the focus back to reciprocity and balance before you become completely drained.

Social relationships are always complex, but, again, negotiating this complexity is a large part of what drove the evolutionary development of our big brain in the first place. We simply have to remember the bumper-sticker wisdom to take life "one day at a time." Alcoholics Anonymous also has another apt saying: "The road to recovery is always under construction." The same is true for the road to healthy social connections.

When you are trying to hold on to a valuable relationship, three points discussed in earlier chapters are especially important to keep in mind:

- *Loneliness can make us demanding.* It is typical for both people in a relationship to believe they are doing at least their fair share. After all, they see everything they do and forgo for the relationship, and they do not see everything the other person does or forgoes. In the most successful couples, both partners understand that their perceptions of what they contribute are biased, so to make sure they are properly tending to the relationship, they try to do *more* than their share. After a quarrel, members of the happiest couples also give each other the benefit of the doubt. After their spat, they do nice things for each other to reinstate their love and trust.[8] They do not match negative with negative in a downward tit for tat. Instead, each will inject a positive comment or gesture into the loop to reset the standard upward. By offering overtly cooperative behavior, they elicit further cooperation from their partner.

- *Loneliness can make us critical.* Members of couples that are high in social well-being find ways to idealize their partners, sustaining what are called positive illusions.[9] (This fictive element is why they call romance *romance,* the original term for popular narratives.) A thirteen-year study of marriages showed that idealization of the partner not only helps sustain love, it also lessens the likelihood of divorce.[10] Idealizing the partner does not mean overlooking deceit or abuse or other serious issues, but it may mean focusing on the still beautiful smile instead of the cellulite or the thinning hair, or recognizing the way he shows his love by scraping the ice off your car, even if he could be better at using words to express his feelings. The executive brain gives us a great deal of control over what we choose to emphasize—but only if we keep the fear-induced disruptions of loneliness out of the way.

- *Loneliness can make us behave passively and withdraw.* People in happy relationships take active steps to capitalize on the positive events that occur in their daily lives.[11] When people tell friends or loved ones about a pleasant moment from their day, sharing the experience provides positive affect and a greater sense of well-being in addition to the benefits already accrued from the event itself. The surprising finding is that having a romantic partner who reacts actively and constructively to your good fortune is actually more conducive to a happy marriage than having a partner who can soothe you in the bad times. So when your partner offers you that breath of fresh air or that glimpse of a silver lining, don't let it pass by. Reach out to share the enjoyment; both of you will benefit.

We don't always need words to express the positive emotions that we feel toward others and want to feel from them in return. Make full use of the powerful effects of oxytocin. Disputes often grow and fester, spawning coldness and resentment that engender additional disputes which reinforce the coldness and resentment. Even when we can't find the right words, many times we can stop this negative feedback loop just by silently taking someone's hand, or, corny as it may sound, by giving them a hug.

For those who care about someone who seems to suffer from feelings of social isolation, whether in personal life or in the world of work, two further reminders may be useful:

- *Be aware of the underlying reality.* Understand that much of your friend or loved one's disagreeable behavior may be the result of fight-or-flight responses to a sense of being unsafe in the world, and that you can't win by arguing. Trying to step outside the frame and address the distorted cognition itself is more promising, but still difficult, and sometimes that too will only reinforce resistance. The most effective approach often is to directly address the person's most basic emotions, which include dejection and fear. Remember that we humans often use words and logic merely to rationalize our primitive emotions and prior expectations.

- *Do what you can to make the lonely person feel safe.* It makes no difference that a perception results from a feeling of threat that cannot be logically justified—a feeling simply is what it is. Feeling unsafe often stems from a deep, underlying feeling of rejection, so first and foremost, do what you can to provide a sense of safe enclosure. If you are dealing with a family member or an intimate partner, try to demonstrate that your love is rock solid. You may feel as if you are surrendering in today's battle, but you may gain victory in the long run.

Perfect friendships are impossible, but by reaching out beyond ourselves we can achieve the next-best thing—social connection that is rich and satisfying, even as it demands effort and forbearance from us. Ultimately, the secret lies in using our most distinctly human capacities in order to find solutions that benefit everyone involved. These are solutions to which each partner contributes, that neither could have anticipated, and that exceed what either individual could have accomplished on his or her own.

Many of the same principles derived from the study of loneliness that we can apply to improving our private lives are also applicable to the larger social environment. In that wider world, as we will see, the power of social connection can be a vital force for change.

CHAPTER FOURTEEN

the power of social connection

In 1985, when researchers asked a cross-section of the American people, "How many confidants do you have?" the most common response to the question was three. In 2004, when researchers asked again, the most common response—made by twenty-five percent of the respondents—was none. One-quarter of these twenty-first-century Americans said they had no one at all with whom to talk openly and intimately.[1]

Also published in 2004, a joint study by the World Health Organization and researchers from Harvard University found that almost ten percent of Americans suffer from depression or bipolar disorder. They also found that binge eating and drinking are up, and that our children are medicated for depression and attention deficit disorder to an alarming degree.[2]

When UNICEF surveyed twenty-one wealthy nations, the United States came in second to last in terms of the welfare of its children, with only the United Kingdom faring worse. The United States had the very worst record in terms of infant mortality rates, and second to worst in terms of exposure to violence and bullying, chaotic family structure, and troubled relationships with family and friends. Respondents to the survey from across the United States say that their families no longer have meals together. Children say that they don't spend time talking to their parents, and that they generally don't find their peers kind and helpful.[3]

For citizens of the twenty-first century, "the way things used to be"—being bound to your village, marrying someone chosen by your family, and otherwise doing whatever your priest or your parents or your tribal elders tell you to—is not a life plan with much appeal. However, the dismal statistics above suggest that our society may have gone overboard in its emphasis on standing alone. We pay the price, not just in terms of our mental and physical health, but in terms of the strain on social cohesion and sustainable economic progress. The corollary to being "obligatorily gregarious" is being interdependent. "Independence," the biologist Lynn Margulis reminds us, "is a political, not a scientific term."[4]

And yet independence is the rallying point for our culture. We have always prized vertical mobility and accepted "horizontal mobility" as the cost of doing business—you go where the opportunities are. By the middle of the twentieth century, however, that swashbuckling independence could be better described as rootlessness. Executive transfers had become a staple of even the most routine and regimented corporate lives, turning managers into a new species of migrant worker. The triumph of the interstate highway system, tract housing, strip development, and the automobile encouraged the creation of interchangeable landscapes, with entire "communities" mass-produced as marketable commodities. Sales people, consultants, and even academics like me became road warriors, racking up the frequent-flyer miles.

Landscapes for Loneliness

In the 1950s the sociologist Robert Weiss began to explore the effect of new working patterns and living patterns on loneliness. He noted that "low population density and the loss of natural daily social gatherings on the porch, the street, or the corner drugstore made sharing experiences and insulating problems more difficult."[5] Residents of transient communities lacked not only long-term relationships with friends and neighbors but the benefits of living close to older generations of their own families.

Weiss's colleague Mark Fried referred to the loneliness of working-class residents of Boston's West End "grieving for a lost home" after their neighborhood was razed for what was then called urban renewal.[6] This was a community of people rich in attachments, both to the place and to one another. Just a few years ago you could get a taste of what the West End had been like by walking through Boston's North End—a chaotic jumble that seemed to operate as an extended family. But now gentrification threatens the established connections in that community as well.

In most industrialized nations, champions of modernism like New York's "master builder" Robert Moses continued until very recently to bulldoze older neighborhoods to run expressways through cities, and urban planners built huge housing projects—"vertical slums"—to warehouse the poor. The apartheid government of South Africa went so far as to destroy a wide swath of Cape Town—a mixed-race area called District Six—precisely *because* of its rich sense of community. The harmony that had flourished among the district's crowded mix of blacks and whites and Asian immigrants gave the lie to the ruling party's agenda of racial separatism.

In the 1960s urbanists like Jane Jacobs launched a counteroffensive. Jacobs's book *The Death and Life of Great American Cities* is her paean to her own "village"—Greenwich Village in New York City. In its pages she extols the vitality of life on a smaller, more compact scale, where people live and work on the same block. She writes about the greater trust and sense of connection, as well as the enriching, serendipitous encounters that result. I can attest to her insight, because my wife and I live in just such an urban village, a cluster of nineteenth-century row houses where neighbors know one another's children and pets and keep up with the progress of one another's plantings beside the doorsteps. My coauthor lives in a small New England town where members of the same families have rubbed shoulders since the 1630s, and where lobstermen and lawyers go to the same parties. As kids, each of us had bounced around in various places in the southwest, and so as adults we each made a choice about where to put down roots quite deliberately. But even though we have been lucky enough to find pockets where community flourishes, elsewhere the war on human scale and human bonds continues.

Bowling Alone

In a book entitled *Bowling Alone: The Collapse and Revival of American Community*, the political scientist Robert Putnam explores the implications of our atomized culture in terms of lost "social capital," a phrase he uses to refer to the reciprocity, cooperation, and collective goodwill derived from connection with the larger community. In recent years, Putnam notes, participation in all forms of civic engagement has sharply declined, from voter turnout to bridge clubs, from volunteer fire departments to marching bands, from alumni organizations to bowling leagues.

"Civic virtue is most powerful when embedded in a dense network of reciprocal social relations," he writes. "A society of many virtuous but isolated individuals is not necessarily rich in social capital." But many affluent towns no longer have housing that fits the budgets of nurses, teachers, and police officers—the kinds of workers who help stitch a community together. When vital services depend entirely on civic-mindedness, as is the case with volunteer fire departments, the problem is even more acute. Investment bankers may contribute mightily to a community's tax rolls, but high earners with ridiculously demanding jobs tend to be less eager than others to commit themselves to come running when their neighbor's house is on fire.[7]

Whether on the level of civic engagement or more intimate connection, the march toward atomization continues. Feelings of isolation engender depression and hostility and impair self-regulation. Nonetheless, many political leaders cut funds for community-building in favor of building larger prisons for those whose lack of self-regulation makes them hostile and out of control. The data tell us that loneliness seriously accelerates age-related declines in health and well-being, yet the idea of promoting connection is rarely discussed alongside the heated issues of the cost of pharmaceuticals and other medical interventions necessary to deal with an increasingly lonely, isolated, and aging population.

At long last our national consciousness may be awakening to the idea that protecting our natural environment, including the global climate, is not some harebrained idea from the Sixties. Given the statistical impact of loneliness, if its effects were caused by an impurity in our air or water, perhaps now there would be congressional hearings on how to reduce it. Perhaps we can hope for a similar awakening to the idea, grounded in rigorous science, that restoring bonds among people can be a cost-effective and practical point of leverage for solving some of our most pressing social problems, not the least of which is the looming crisis in health care and eldercare.

But given the world as it is today, what can we do to cope?

Lonesome No More

In the tougher neighborhoods of urban areas, today's disaffected youth respond to the dangers of being alone by signing up with the Crips or the Latin Kings. In the hip coastal enclaves, the more affluent young try to create the kinds of surrogate families they see on reruns of *Seinfeld* or *Friends*. Couples with children consciously pursue togetherness, trying to combat the centrifugal force exerted by media that divert the attention of each family member into a separate room, or at least into a different portion of cyberspace. Yet absent a supportive, integrated community, or some natural bond such as shared work, these attempts sometimes appear forced. A "family first" preoccupation sustains life in kid-centric suburbs, but it deprives the adults of a broader range of social supports. As Weiss noted years ago, itinerant nuclear families, relocating in and out of faceless suburbs, necessarily focus inward, a situation that places intense emotional demands on family members to be "all" for one another. Stephanie Coontz, a sociologist and the author of *Marriage: A History*, decries the growing numbers of people who now depend on their spouse as their one and only source of companionship.[8] Perhaps it should not be surprising that so many more Americans today than twenty years ago have no confidants. To whom can you speak in confidence when your most agonizing personal issues might have to do with your spouse?

In 1976 the novelist Kurt Vonnegut told the story of Wilbur Swain, a pediatrician who runs for president of the United States with the slogan "Lonesome No More." Swain's winning platform consists of a plan to create artificial families—designated by new, totemic middle names—so that every citizen would have ten thousand brothers and sisters.[9]

Twenty years before this fictional proposal for solving the problem of isolation, the Reverend Robert H. Schuller began a ministry in Orange County, California, that would have made Wilbur Swain proud. At first preaching from atop the snack bar at a drive-in movie theater on Sunday mornings, Schuller tailored his message to meet the needs of socially disconnected transplants from the Midwest. Five decades later, his pulpit (now usually filled by his son Robert A. Schuller) is *The Hour of Power*, a television program broadcast worldwide from the multimillion-dollar Crystal Cathedral, built on a fifteen-acre campus that draws visitors from every continent. The simple message that carried this ministry from a dusty drive-in to a global media empire is summed up in the signature line used in every broadcast: "God loves you and so do I."[10]

In recent years the attempt to form more intense social bonds has helped drive the explosive growth of new megachurches, replicating Schuller's model, from Kansas to Korea. In the modern exurban areas patterned on Orange County, large numbers of people appear more desperate than ever for a sense of community and meaning—and if God can be a part of it, adding even greater meaning and structure, so much the better. But a focus on the human need for connection and social support was a central part of Christianity long before its adherents came to be called Christians. The same concern is a key element of Confucianism, Buddhism, Islam, and Judaism—all the faiths that have large numbers of adherents.

In the early days of the Jesus movement, sects like the Gnostics who were mystical and inner-directed quickly faded out of existence. The type of Christianity that went on to become the primary structural element of the Western world focused on a simple message of self-esteem—"the kingdom of God is within you"—combined with communal meals and even communal living. Its streamlined theology set aside the complex cleansing rituals of Judaism, and it presented evil less in mystical terms and more as a question of the behavior of one person toward another. The church that survived and prospered extended the basic ethics of the Hebrew tradition—already a strong source of social support—explicitly into the individual's inner life, creating prohibitions against mere thoughts that were harmful to social connection: anger, hatred, misdirected lust. It dispensed with the temple in Jerusalem as the center of religious life, but maintained rituals to sanctify the basic elements of ordinary human existence: reproduction (marriage), birth (baptism), illness (anointment), and death (last rites). By way of these ceremonies it provided guidelines for social connection throughout the life cycle, making this universal church a practical social convention: It offered self-worth, it buried the dead, and it provided for the poor. Like Judaism, Islam, Confucianism, and Buddhism, Christianity regulated all social transactions within the community, ranging from relationships within marriage and the family to standards for conducting business and dealing with neighbors.

Two thousand years along this path, we find Christian leaders like Joel Osteen, a member of a second generation of megapreachers, acquiring and refitting Houston's Compaq Center, a professional basketball arena, in order to accommodate his growing congregation. We find Rick Warren, pastor of his own megachurch, reaching millions with the explicit chord of social connection in his book *The Purpose Driven Life.* One of the biggest bestsellers in recent years, the book outlines God's supposed five directives for each of us. This is number two on the list: "We were formed for God's family, so your second purpose is to enjoy real fellowship."

The Christian megapreachers and their megachurches have been so successful that even some Jewish leaders, specifically a group called Synagogue 3000, have rigorously studied their methods, sending representatives to attend seminars on congregation-building at Rick Warren's Saddleback Church.[11] But the key element of their success seems to be that these new churches, set among the sprawling office parks, "big box" shopping destinations, and "planned communities" of exurbia, reflect the basic human need to gather, connect, and belong. In doing so they adventitiously address human loneliness in each of the three dimensions—intimate, relational, and collective. From education, to dating services, to daycare, to psychological and marital counseling, to basketball tournaments, they provide one-stop shopping for human connection in many different forms.[12]

The growth of the megachurches suggests that they serve a need, but they do so in ways that those of other faiths or those who are not overtly religious can find troubling. This is community with a specific worldview and a specific agenda, not community that can provide connection for anyone and everyone on the basis of simple, shared humanity. And yet in many parts of North America no other institutions exist to combat the oppressive feeling of being both physically and spiritually isolated.

In a similar way, a younger generation is finding connection of a sort in "virtual worlds": massive multiplayer online communities with names like Second Life, There, and Active Worlds. These sites allow users to create avatars—physical representations of themselves on the screen—who then mix and mingle, buy real estate, furnish homes and other meeting places, and otherwise carry on the routine business of life, only in cyberspace. These meta-universes, or "metaverses," are not games, exactly, because the participants have no specific objective, no way to "win." The point of this online activity, as is the case with "theology lite" megachurches in the exurbs, is to experience a sense of community.

Global Disconnection

In many parts of the world, older societies are rushing to embrace the American commodity culture and the casual disregard for social bonds that gave rise to the exurb's anomie.

In China a society built on Confucian regard for the collective has been suddenly thrust into the aggressive individualism of capitalism. The *New York Times* has reported on the growth of "instant" cities in China's heartland that would make the transformation of Phoenix or Las Vegas seem sluggish. In Yinchuan, the capital of the Ningxia region, officials are spending over a billion dollars a year to create a huge government complex, a five-star hotel, and a residential compound for entrepreneurs, in the hope that the infrastructure will attract private real estate development. Dozens of other provincial towns have the same aspiration, hoping to turn peasant villagers into citizens of the global economy overnight. Lu Dadao, a Beijing expert on urban planning, told the *Times* reporter Jim Yardley: "They want it to happen fast, and they want it to be big. They have all taken up urbanization without considering what the natural speed of it should be." In terms of health and well-being, science tells us that there are unintended negative consequences when, as Walter Lippmann put it a century ago, "we have changed our environment more quickly than we know how to change ourselves."[13]

Here in the United States, progressive architects and developers have heeded Jane Jacobs's call to take the imperatives of social connection more seriously. They try to replicate, in new communities such as Celebration, Florida, the physical aspects of small-town life—clustered housing, sidewalks, front porches for sitting—that facilitate social connection. Other communities, such as Treetops in Easthampton, Massachusetts, try to reintegrate older and younger people in a single living arrangement. In the United Kingdom, the Prince of Wales has championed attempts to mirror the traditional English village in contemporary housing. Unfortunately, these efforts remain noble islands of experimentation in a sea of sprawl. All over the world, globalization now threatens to make the anonymity and interchangeability of American places, if not the norm, at least distressingly familiar.

But the battle is not over. A landscape built for disconnection simply makes it even more urgent to work consciously and deliberately to build stronger human bonds at every opportunity, in every day-to-day exchange. It places an even greater premium on the kind of reaching out, as well as the "random acts of kindness," that we discussed earlier. It also means that in our intimate relationships we need to be aware of what we're up against. A commodity culture can foster a "consumer" mentality, encouraging us to apply concepts such as "trading up" and optimizing value in our romantic partnerships. As the psychologist Elaine Hatfield told a reporter, "People demand so much more. I don't think that's bad; it's just a different problem. In the old days, there was not the notion that you were entitled to personal happiness. Now, people want it all: good looks, money, intelligence, status."[14] Perhaps the apotheosis of this kind of thinking was a billboard in Chicago (quickly removed after a flurry of protests) that showed two beautiful torsos, one male, one female. The advertising copy, promoting an attorney's services, said, "Life is short; get a divorce."

Most people don't go to such extremes of crassness, but unrealistic (and superficial) expectations do lead to disappointment. As Hatfield commented, "I think that's why women end up with pets and guys end up with computers."[15]

Working with What We've Got

The kinds of connections—pets, computers—we substitute for human contact are called "parasocial relationships." You can form a parasocial relationship with television characters, with people you "meet" online, or with your Yorkshire terrier. Is this an effective way to fill the void when connection with other humans, face to face, is thwarted?

The Greeks, specifically the pre-Socratic philosopher and poet Xenophanes, used the term "anthropomorphism" (combining *anthropos*, meaning human, and *morphe*, meaning form) to describe the projection of specifically human attributes onto nonhuman entities. Increasing the strength of anthropomorphic beliefs appears to be a useful tactic for coping with loneliness, divorce, widowhood, or merely being single.[16] Pet owners project all sorts of human attributes onto their animal companions, and elderly people who have pets appear to be buffered somewhat from the negative impact of stressful life events. They visit their doctors less often than do their petless age-mates. Individuals diagnosed with AIDS are less likely to become depressed if they own a pet. In circumstances in which one is going to be evaluated, the presence of one's pet can actually do more to reduce anxiety and psychophysiological responses to stress than the presence of one's spouse.[17]

One of the lessons of Hurricane Katrina was that pet owners were so committed that many were willing to risk their lives to remain in the city to care for their animals. Was it the sense of being left alone with the elements—in a sense, rejected by those who fled the storm—that made their attachment so strong? Studies show that rejection by other humans can increase the tendency to anthropomorphize one's pet.[18] Perhaps many of these economically deprived people had felt rejected all along. All we know for sure is that the number of people who had to be forcibly removed from their homes (and were forced to leave their animals behind) led to the passage of the Pets Evacuation and Transportation Standards Act, signed into law in October 2006. This law requires local and state emergency-preparedness authorities to include in their evacuation plans ways to accommodate household pets and service animals in case of a disaster. It also authorizes federal funds to states to help establish pet-friendly emergency shelters.

In the movie *Castaway*, when the character played by Tom Hanks is stranded on a desert island, he forms an intense relationship with a volleyball named Wilson. Similarly, the retired academic who told me about her new social landscape once she returned to the Midwest also described how she had coped with loneliness during a semester spent doing research in Paris, an ocean away from her husband and her cats. She could see the Eiffel Tower from her bedroom window. In letters and phone calls to her husband she called the Tower her "pet." Each evening as she turned out the light she would say goodnight to it.

Social rejection, even in fleeting episodes, can also increase people's belief in anthropomorphized supernatural agents.[19] The need to compensate for their late partners' physical absence often leads widows or widowers to carry on "two-way" conversations with them. Loss of a husband or a wife may increase the survivor's belief in devils and gremlins as well as kindly ghosts or angels, indicating that parasocial connection is not simply an attempt to soothe the mind with positive images and repair negative mood.[20] But whether it's a god, a devil, an animal, a machine ("Old Betsy"), a landmark, or a piece of cast-off sports equipment, the anthropomorphized being becomes a social surrogate, and the same neural systems that are activated when we make judgments about other humans are activated when we assess these parasocial relationships.[21]

For an experiment, my colleagues Adam Waytz, Nick Epley, and I collected photographs taken by the Hubble Telescope of dramatic celestial bodies such as starfish nebulae and "Bok" globules and showed them to people sitting on the lakefront and in city parks in Chicago. After the respondents viewed each photograph, we asked them a series of questions. Some were simple queries about function and appearance, but some delved into the realm of human attributes. Were these structures simply clouds of gasses floating in space, or did they have certain human characteristics? Was the celestial object, for instance, moving from here to there with a purpose? At the end of the survey we measured the respondents' levels of loneliness. Responses from those who were high and low in loneliness were very similar except in one regard. The lonelier respondents showed a stronger tendency to see the celestial objects as having human characteristics, even as acting on the basis of lessons learned from past experience. Like our ancient ancestors who first named the constellations and gave them life stories, our lonely Chicagoans had anthropomorphized the objects we see in the distant sky.

Partners We Can't See

Our parasocial relationships follow certain patterns based on aspects of our human relationships. People with insecure, anxious attachment styles are more likely than those with secure attachment styles to form perceived social bonds with television characters. They are also more likely than those with secure attachment styles to report an intensification of religious belief over a given time period, including sudden religious conversions later in life.[22]

In a *Newsweek* poll of religious beliefs in America, forty percent of respondents indicated that they felt closer to God when praying alone, while only two percent indicated feeling closer to God when praying with others.[23] Nuns, monks, and mystics apply this intensifying effect of isolation as a positive when they remove themselves from other humans in order to "feel the presence of God" more powerfully. Again, the feeling of isolation promotes not only the drive to connect, but the intensity of anthropomorphism.

Many proponents of technology tell us that computer-mediated social encounters will fill the void left by the decline of community in the real world. The Lions Club, the Masons, the barbershop quartet, or the bowling league may be fading away, but that's okay, these enthusiasts tell us, because everyone is busy texting each other or "connecting" in chat rooms. Email, however, is what communication theorists call a single-stranded interaction—words on a screen devoid of any other physical texture. Studies have shown that the richer the medium—the more physicality it has—the more it fosters social cohesion. This may be why, for those who do choose to connect electronically, multiplayer sites like Second Life are becoming popular meeting places. These virtual communities are at least enriched to the extent that each participant has an avatar, an animated physical representation that appears on the screen. Participants also build (or pay web designers to build for them) well-appointed meeting places. Thus the real people sitting at home in front of their computer screens can come together "avatar to avatar" in virtual bars and clubhouses and react to one another with animated gestures and facial expressions.

And yet, most face-to-face encounters in real life allow us to communicate through even more subliminal cues—body chemistry, body language, action semantics, mimicry—in addition to words and gestures. Once again, the mind that seeks to connect is first about the body, and leaving the body behind can make human connections less satisfying.

When being physically together is not possible, we try to satisfy our yearnings by speaking briefly on the telephone, sending an instant message, or gazing at a loved one's photograph, practices that have been called "social snacking"—but a snack is not a meal.[24] A military friend of mine described the problem created by the introduction of satellite phones to the modern war zone. During his tours in Afghanistan and Iraq, he and his comrades were eager at first for any chance to call home. They quickly learned, however, that the sudden juxtaposition of two such very different worlds—the battlefield and the family room—was not just unsatisfactory but emotionally upsetting, both to the men in the field and to the wives and children at home. He said you could always tell who had just called home by his empty, "thousand-yard stare." The abstracted nature of electronic communication—the absence of physical context and forms of connection—may account in part for the finding that increased Internet use can increase social isolation as well as depression when it replaces more tangible forms of human contact.[25]

Again, forming connections with pets or online friends or even God is a noble attempt by an obligatorily gregarious creature to satisfy a compelling need. But surrogates can never make up completely for the absence of the real thing. In a culture built around disconnection, the better move is to work that much harder to reach out to those with whom we share even the most superficial contact in the everyday world.

Gatherings

As an obligatorily gregarious species, we humans have a need not just to belong in an abstract sense but to actually get together. Congregating physically may actually play a role in an association found between religious observance and decreased morbidity and mortality. The sociologists Lynda H. Powell, Leila Shahabi, and Carl E. Thoresen conducted a meta-analysis of the extensive literature on religion and health, exploring nine different hypotheses that might account for the purportedly positive effects. Do religious people live longer and healthier lives because of the more conservative and healthful lifestyle that religion promotes? Is it the power of prayer? Or is it something about spirituality in itself that is affecting us at the cellular level?[26]

After sorting through mountains of data, the three authors found no association between depth of spiritual feeling and health. Instead, what they found was a strong, consistent, prospective, and often graded reduction of mortality linked to individuals who actually attend religious services. In other words, people who regularly went to church or synagogue lived longer than those in similar situations who did not. In some studies there is even a "dose effect," meaning that those who go to church more than once a week enjoy even better health than those who attend only once a week. Overall, the reduction in mortality attributable to churchgoing is twenty-five percent—a huge amount in epidemiological studies—even after discounting other effects, such as the fact that, yes, being religious generally leads to a more healthful lifestyle.

The authors cite the possibility that those who are sufficiently devout to attend services at least once a week may also practice calming techniques associated with religion, practices such as meditating or praying or saying the rosary. But as I mentioned earlier, people smile more when watching a film in a friend's presence, even when they report that their actual enjoyment of the film is no greater. We are social mammals, and, all other things being equal, congregating among our fellows feels good, and that good feeling undoubtedly amplifies the benefits of other positive experiences.

Weekly attendance at the Rotary Club may also be good for you, but the findings by Powell and her colleagues indicate that there may be something unique about regular attendance at *religious* gatherings. Church attendance often has the added benefit of reinforcing family connections and providing trustworthy interactions with friends. Religions also tend to focus on helping others, rather than on being helped. This altruistic focus fosters feelings of self-worth and control while reducing feelings of depression. Attendance at religious services also affords social modeling—seeing others committed to compassionate helping, as well as prayer and meditation—that reinforces various positives, including a healthier lifestyle. The sense of community, the time spent in the presence of good friends, the reinforcement of the intimate connections of marriage and family, may all contribute to the boost in well-being. And yet there may be something operating that is more powerful still.

My colleague Nick Epley has found that people attribute to other people attitudes that are fairly similar to their own, but that believers attribute to God attitudes that are *uniquely* similar to their own. A country song called "Me and God," written by Josh Turner, captures this idea with a line about the singer and God being like "two peas in a pod." Anne Lamott's variation on the same theme is, "You can safely assume you've created God in your image when it turns out God hates all the same people you do."[27]

From the perspective of cognitive science, God can be a distinctive psychological projection in which people assign their own beliefs and prejudices to the author of the universe. The projection provides an intimacy, an affirmation of self—at least of an idealized self—that is not as evident or as powerful in any other parasocial relationship. God is uniquely self-affirming, because, in the eyes of the believer, God is uniquely "me."

As social beings with a DNA-based interest in the future, we are driven to look beyond ourselves not just for connection but for meaning. The "selfish gene" led to a social brain. That social brain reinforced the aversive response to loneliness that reinforced human connection, thus improving our chances of survival, and thus the survival of our genes. Eventually, in a continuing progression, the same shaping forces of natural selection gave rise to the Third Adaptation, which involves seeing our genes' long-term self-interest in the context of reciprocity and interdependence with other members of our species. This drive for meaning appears to have endowed us with a biological need to be linked with something greater than ourselves. It is only through some ultimate sense of connection that we can face our own mortality without despair. Knowing that our biological existence is transient, we yearn for the transcendent experience described by the astronaut Edgar Mitchell when he looked back at the earth from the moon and perceived that the universe was "intelligent, loving, harmonious."[28] Just as finding social connection is good for us, finding that transcendent something appears to be very good for us, whether it is a belief in a deity or a belief in the community of science. Of course, the danger in transcendent feelings is when the sense that I am at one with the universe becomes corrupted by the sense that the universe is at one with me. Too often throughout human history, when a strong parasocial relationship with its projection of self has replaced a respectful sense of awe, the feeling that "God is on my side" has led to the conclusion that "everyone else must do as I say." This is still a source of human misery wherever there is not a firm separation between private faith and public life.

Once more, then, we come back to the urgency, no matter what our religious beliefs or lack thereof, of satisfying our psychological and physiological needs for connection, including our need for transcendent meaning, through contact with and concern for other people, by being open to and accepting of others, and by "feeding others," in the everyday here and now.

Choosing Our Future

Throughout this book I have emphasized that much of our social reality is something over which we can exercise a certain degree of control. Even with regard to the forces that are outside our control, the way we interpret them, cope with them, and act in response to them can have dramatic effects on our future. This operates at a societal level as well as at an individual level. We as individuals and as groups can choose to make the most of the Third Adaptation—seeking solutions through committed actions that benefit the greater good well beyond ourselves or our tribe—or we can stay back with the chimps in remaining more narrowly focused and self-interested.

My hope is that understanding the biology of loneliness will allow us to see that ethical, humane behavior is a prescription for greater well-being, even economic well-being. This is a message worth heeding, because even in strict dollars-and-cents terms, the cost of social isolation is staggering.

The health consequences of loneliness that I described in Chapter Six carry a heavy price tag in and of themselves. But when we consider the degree to which growing older can contribute to loneliness, and the rate at which our population is aging, it is clear that we need to rethink many of our priorities.

The United States has experienced an enormous growth in wealth since the 1970s, but that rise in income has increasingly benefited those already at the top. Those in the middle or at the bottom have seen their economic condition remain the same or deteriorate. In just the past few years, the size of the disparity has exploded. From 1990 to 2004 the income of the poorest ninety percent of Americans grew by only two percent. During the same period, the income of the richest one percent grew by fifty-seven percent, and the income of the top one-tenth of one percent—the superrich—grew by eighty-five percent![29]

Economic growth is accelerating elsewhere around the globe, particularly in China and India. The past twenty years have seen rapid economic expansion in Russia, and yet, today, Russians are dying younger than they were under Soviet oppression. Since the 1980s their longevity has declined by forty percent, putting them on a par with Bangladesh.[30] A rising tide can indeed lift a variety of boats, but in a culture of social isolates, atomized by social and economic upheaval and separated by vast inequalities, it can also cause millions to drown.

The Economics of Isolation

Money appears to have a positive impact on people's motivation, but a negative impact on their behavior toward others. There are data to suggest that merely having money on the periphery of consciousness is sufficient to skew us away from prosocial behavior. The psychologist Kathleen Vohs and her colleagues did a series of nine experiments that primed certain participants with thoughts of money. Some were asked to unjumble phrases that included mental images such as "high a salary desk paying" while the controls unscrambled neutral phrases such as "cold it desk outside is." Other groups worked on computers which, after a few minutes, showed a screen saver that was either fish sparkling underwater or currency sparkling under water. In all nine tests, those who were given the subtle suggestions of money were not only less likely to ask for help, but also less likely to help others. When a lab assistant staged an accident by dropping a box of pencils, those primed with thoughts of money picked up far fewer. When the experimenter asked for help coding data sheets, the primed participants donated roughly half as much time as the non-primed participants did. When asked to choose activities from a list, the primed were far more likely to opt to work and play alone. And when they were given a chance to set up chairs for an interview, they chose to put greater physical distance between themselves and other people.[31]

In a similar vein, various studies have attempted to correlate income inequality with health statistics in each of the fifty U.S. states.[32] Bruce Kennedy and his colleagues developed something called the Robin Hood Index, referring to the amount of wealth that would have to be redistributed to attain an equal distribution. They found that an increase of one percent in this measure of inequality was associated with additional mortality of 21.7 deaths per 100,000 people. Their analysis isolated three possible explanations for this finding: (1) relative deprivation (if wealthy people own three houses each, that increases price demand on all housing); (2) underinvestment in human capital (less spending on education and health care for the population at large); and (3) erosion of social cohesion, meaning a lack of trust and an increase in feelings of social isolation.

Providing support for the importance of this third explanation, other researchers have found that socially integrated societies have lower rates of crime and mortality and a better quality of life overall.[33] In thirty-nine states, citizens were asked to list their group memberships. An increase in average per capita memberships by one unit was correlated with a decrease in mortality of 66.8 per

100,000. Lower levels of trust within the local culture were associated with higher rates of mortality for every cause of death, including cardiovascular disease, cancer, and infant mortality. One interpretation of such data: Social isolation, including social fragmentation, can kill.

Henry Melvill wrote of our causes returning to us as effects; complexity theorists have their Butterfly Effect. Whether we think in terms of "sympathetic threads" or of autonomous agents acting in a complex system, the fact remains that individual behaviors created both the peace and beauty of Middlebury, Vermont, and the tribal warfare of the Sunni triangle. Of course vast economic, political, and cultural forces are also at play, but ultimately, human beings shape their environment through individual, iterative behaviors. As a free agent within such a system, each of us has a certain degree of power, through our individual actions, to continuously adjust the social environment toward something slightly better or something slightly worse. Simply driving to work, you have the option of extending courtesy or road rage. And sooner or later you, or your spouse, or your children, will encounter the same fellow citizens who have been either goaded by your anger or inspired to their own acts of generosity by the example of your beneficence.

In the winning solution to Robert Axelrod's Prisoner's Dilemma tournament (described in Chapter Four), the computer program called Tit for Tat, we saw the benefits of cooperation as the default mode—albeit with sanctions held at the ready. In his book *The Evolution of Cooperation*, Axelrod offers an example of how similarly beneficial social compacts can self-organize in the real world. In World War I, in certain districts in which they were left in place long enough, soldiers who faced one another in the front-line trenches evolved an ad hoc policy of "live and let live." On their own and in direct defiance of their officers, the enlisted men on the front lines said, in effect, "What's the point in my shooting one of your buddies, if all it means is that you will retaliate and shoot me or one of my buddies?" Neither side was capable of launching a decisive attack for which wearing down the other side might yield an advantage. So the men acted spontaneously. The first step was an informal cease-fire at suppertime. Eventually, this extended round the clock, meaning that they fired only the minimum number of rounds necessary to placate their officers. Even at that, the men on both sides purposely fired off target. Snipers put on displays of marksmanship—firing patterns into the opposing ramparts—to demonstrate the degree to which they were holding back, thus encouraging the other side to appreciate their restraint and to follow suit. Officers from the command center had to offset this spontaneous reciprocity by continually rotating units along the front.[34]

As in Axelrod's iterated Prisoner's Dilemma tournament, in which each round of the competition consisted of hundreds of moves, this kind of socially benign strategy is the best way to go only if you are going to be dealing with the same people over time. But the friends, associates, neighbors, and opponents we confront today—unless we anticipate a life lived constantly on the lam—are more or less the same friends, associates, neighbors, and opponents we will confront tomorrow, and the day after that. This is true both on the level of our immediate community and on the level of the community of nations.

John Donne, the seventeenth-century poet who was also an Anglican priest, wrote: "Any man's death diminishes me, because I am involved in mankind." Many people (erroneously) consider Charles Darwin to be the antithesis of religious thinking, yet coming from his very different perspective, he arrived at a very similar formulation:

> As man advances in civilization, and small tribes are united in the larger communities, the simplest reason would tell each individual that he ought to extend his social instincts and sympathies to all the members of the same nation, though personally unknown to him. This point being once reached there is only an artificial barrier to prevent his sympathies extending to the men of all nations and races.[35]

Perhaps this similarity in thinking is evidence that there is a deeper truth about the human species waiting to be discovered by each of us. As evolutionary psychology and social neuroscience converge, more and more the scientific findings align with the most basic ethical teaching of the most enduring systems of belief, what we call the Golden Rule. It may be that variations on the command "Do unto others as you would have them do unto you" appear in so many different traditions—from the Tao of ancient China, to the law of Moses, to the Sermon on the Mount, to the coldly rational philosophy of Emmanuel Kant—because that command was, in a sense, written by the hand of natural selection.[36]

The scientific data show that social cooperation is the most adaptive option, but, as we know all too well from what we often see around us, it is only one option among many. Which puts a premium on another behavioral admonition. Whether we are trying to break free from individual loneliness or trying to improve the world, we will do well to follow Gandhi's advice to "be the change you want to see."

But once more, societies do not achieve beneficial—and sustainable—levels of social connection and social harmony simply by offering warm hugs and unconditional love. According to the science of complexity, even self-organizing systems need a few simple rules. All civilizations have formal as well as informal rules to promote the adoption of adaptive behavior, including taboos, norms, moral codes, and laws. And this process sometimes requires the kind of "altruistic punishment" discussed in Chapter Eleven, meaning the enforcement of sanctions against others sometimes at a cost to the self. Even minor infractions matter because they set a negative tone that cascades into progressively more negative behavior. If it seems okay to throw trash here, more people will throw trash. If we think everyone cheats on his or her taxes, we are more likely to cheat. If we think everyone is paying his or her fair share, we are more likely to pay what we owe. If there is no social stigma attached to teenage drinking, more teenagers will drink. Without the coregulatory function of social disapprobation, "things fall apart," as Yeats told us, and "the centre cannot hold."

Robert Putnam addresses "social capital" as a societal good. We address it as a personal and collective necessity, and as a major issue of personal, societal, and public health. Civic engagement is the chunk of ice we see floating above the surface; below the water line lurks the much deeper issue of individual feelings of isolation. If civic engagement is to contribute substantially to assuaging the problem of loneliness, then it cannot be something merely akin to networking at a trade show. What individuals need is meaningful connection, not superficial glad-handing.

As individuals, and as a society, we have everything to gain, and everything to lose, in how well or how poorly we manage our need for human connection. With new patterns of immigration changing established cultures throughout the world, the importance of transcending tribalism to find common ground has never been greater. We need to remember not only the ways in which loneliness heightens our threat surveillance and impairs our cognitive abilities, but also the ways in which the warmth of genuine connection frees our minds to focus on whatever challenges lie before us. Both as individuals and as a society, feelings of social isolation deprive us of vast reservoirs of creativity and energy. Connection adds more water to the well that nourishes our human potential.

Coming from the religious tradition of John Donne, C. S. Lewis wrote: "We are born helpless. As soon as we are fully conscious we discover loneliness. We need others physically, emotionally, intellectually; we need them if we are to know anything, even ourselves."

Coming from the scientific tradition of Charles Darwin, E. O. Wilson wrote: "We are obliged by the deepest drives of the human spirit to make ourselves more than animated dust. We must have a story to tell about where we came from, and why we are here."[37]

notes

CHAPTER ONE: *Lonely in a Social World*

1. E. Berscheid, "Interpersonal attraction," in G. Lindzey and E. Aronson, eds., *The Handbook of Social Psychology* (New York: Random House, 1985).
2. C. Rubenstein and P. Shaver, *In search of intimacy* (New York: Delacorte, 1982). D. E. Steffick, "Documentation on affective functioning measures in the Health and Retirement Study," Documentation Report no. DR-005 (Ann Arbor: University of Michigan, Survey Research Center, 2000), retrieved February 7, 2006, from *hrsonline.isr.umich.edu/docs/userg/dr-005.pdf.*
3. J. S. House, K. R. Landis, and D. Umberson, "Social relationships and health," *Science* 241 (1988): 540–545.
4. Simply write a down a number between 1 and 4 beside each question in Figure 1 to indicate how often you feel that way. However, note that half the questions are worded in a way that probes what you feel is missing from your life, and the other half are worded in a way that probes what you feel is present. Because both kinds of questions are coming at the same kinds of feelings from opposite directions, we score half the questions with the higher numbers meaning "more often," and half the questions with the higher numbers meaning "less often."

   For the questions marked with asterisks, write down a number to note how you feel according to this ranking:

   1 = Always 2 = Sometimes 3 = Rarely 4 = Never

   For the questions without asterisks, write down a number to note how you feel according to this ranking:

1 = Never 2 = Rarely 3 = Sometimes 4 = Always

Then add up the numbers to find your score. High loneliness is defined as scoring 44 or higher. Low loneliness is defined as scoring less than 28. A score of 33 to 39 represents the middle of the spectrum.

5. J. Bowlby, "Affectional bonds: Their nature and origin," in R. S. Weiss, ed., *Loneliness: The experience of emotional and social isolation* (Cambridge, MA: MIT Press, 1973), 38–52.
6. P. L. Jackson, A. N. Meltzoff, and J. Decety, "How do we perceive the pain of others? A window into the neural processes involved in empathy," *NeuroImage* 24 (2005): 771–779.
7. C. J. Norris, E. E. Chen, D. C. Zhu, S. L. Small, and J. T. Cacioppo, "The interaction of social and emotional processes in the brain," *Journal of Cognitive Neuroscience* 16 (2004): 1818–29.
8. M. Gazzaniga, *The cognitive neurosciences*, 3rd ed. (Cambridge, MA: MIT Press, 2004).
9. Bruskin Associates, "What are Americans afraid of?" *Bruskin Report* 53 (1973): 27.
10. K. D. Williams, *Ostracism: The power of silence* (New York: Guilford, 2001).
11. R. I. M. Dunbar and Suzanne Shultz, "Evolution and the social brain," *Science* 317 (September 7, 2007): 1344–47.
12. I. S. Bernstein, T. P. Gordon, and R. M. Rose, "The interaction of hormones, behavior, and social context in nonhuman primates," in B. B. Svare, ed., *Hormones and aggressive behavior* (New York: Plenum, 1983), 535–561.
13. Alexis M. Stranahan, David Khalil, and Elizabeth Gould, "Social isolation delays the positive effects of running on adult neurogenesis," *Nature Neuroscience* 9, no. 4 (April 2006).
14. R. S. Wilson, K. R. Krueger, S. E. Arnold, J. A. Schneider, J. F. Kelly, L. L. Barnes, Y. Tang, and D. A. Bennett, "Loneliness and risk of Alzheimer's disease," *Archives of General Psychiatry* 64 (2007): 234–240.
15. S. W. Cole, L. C. Hawkley, J. M. Arevalo, C. Y. Sung, R. M. Rose, and J. T. Cacioppo, "Social regulation of gene expression in human leukocytes," *Genome Biology* 8 (2007): R189.
16. J. T. Cacioppo, J. M. Ernst, M. H. Burleson, M. K. McClintock, W. B. Malarkey, L. C. Hawkley, R. B. Kowalewski, A. Paulsen, J. A. Hobson, K. Hugdahl, D. Spiegel, and G. G. Berntson, "Lonely traits and concomitant physiological processes: The MacArthur social neuroscience studies," *International Journal of Psychophysiology* 35 (2000): 143–154.
17. G. R. Semin and J. T. Cacioppo, "Grounding social cognition: Synchronization, coordination, and co-regulation," in G. R. Semin and E. R. Smith, eds., *Embodied grounding: Social, cognitive, affective, and neuroscientific approaches* (New York: Cambridge University Press, in press).

CHAPTER TWO: *Variation, Regulation, and an Elastic Leash*

1. D. Weston, *The political brain* (New York: Public Affairs, 2006).
2. D. I. Boomsma, G. Willemsen, C. V. Dolan, L. C. Hawkley, and J. T. Cacioppo, "Genetic and environmental contributions to loneliness in adults: The Netherlands Twin Register Study," *Behavior Genetics* 35 (2005): 745–752.
3. J. T. Cacioppo, J. M. Ernst, M. H. Burleson, M. K. McClintock, W. B. Malarkey, L. C. Hawkley, R. B. Kowalewski, A. Paulsen, J. A. Hobson, K. Hugdahl, D. Spiegel, and G. G. Berntson, "Lonely traits and concomitant physiological processes: The MacArthur social neuroscience studies," *International Journal of Psychophysiology 35* (2000): 143–154; J. T. Cacioppo and L. C. Hawkley, "Social isolation and health, with an emphasis on underlying mechanisms," *Perspectives in Biology and Medicine* 46 (2003): S39-S52. L. C. Hawkley, R. A. Thisted, and J. T. Cacioppo, "Loneliness predicts reduced physical activity: Cross-sectional and longitudinal analyses," in a symposium entitled "Health behaviors: The relevance of social context and relationship features," Society for Personality and Social Psychology, New Orleans, LA, January 2005, I. Akerlind and J. O. Hornquist, "Loneliness and alcohol abuse: A review of evidence of an interplay," *Social Science and Medicine* 34 (1992): 405–414.
4. J. T. Cacioppo, L. C. Hawkley, G. G. Berntson, J. M. Ernst, A. C. Gibbs, R. Stickgold, and J. A. Hobson, "Lonely days invade the nights: Social modulation of sleep efficiency," *Psychological Science* 13 (2002): 384–387.
5. Cacioppo et al., "Lonely traits and concomitant physiological processes." L. C. Hawkley, C. M. Masi, J. D. Berry, and J. T. Cacioppo, "Loneliness is a unique predictor of age-related differences in systolic blood pressure," *Psychology and Aging* 21 (2006): 152–164. A. Steptoe, N. Owen, S. R. Kunz-Ebrecht, and L. Brydon, "Loneliness and neuroendocrine, cardiovascular, and inflammatory stress responses in middle-aged men and women," *Psychoneuroendocrinology* 29 (2004): 593–611.
6. E. Pennisi, "Why do humans have so few genes?" *Science* 309 (2005): 80.
7. Internal Human Genome Sequencing Consortium, "Finishing the euchromatic sequence of the human genome," *Nature* 431 (2004): 931–945.
8. P. T. Schoenemann, M. J. Sheehan, and D. Glotzer, "Prefrontal white matter volume is disproportionately larger in humans than in other primates," *Nature Neuroscience* 8 (2005): 242–252.

9. G. Roth and U. Dicke, "Evolution of the brain and intelligence," *Trends in Cognitive Sciences* 9 (2005): 250–257.

CHAPTER THREE: *Losing Control*

1. J. T. Cacioppo, J. M. Ernst, M. H. Burleson, M. K. McClintock, W. B. Malarkey, L. C. Hawkley, R. B. Kowalewski, A. Paulsen, J. A. Hobson, K. Hugdahl, D. Spiegel, and G. G. Berntson, "Lonely traits and concomitant physiological processes: The MacArthur social neuroscience studies," *International Journal of Psychophysiology* 35 (2000): 143–154.
2. I. Akerlind and J. O. Hornquist, "Loneliness and alcohol abuse: A review of evidence of an interplay," *Social Science and Medicine* 34 (1992): 405–414. A. W. Stacy, M. D. Newcomb, and P. M. Bentler, "Expectancy in mediational models of cocaine abuse," *Personality and Individual Differences* 19 (1995): 655–667. D. Coric and B. I. Murstein, "Bulimia nervosa: Prevalence and psychological correlates in a college community," *Eating Disorders: The Journal of Treatment and Prevention* 1 (1993): 39–51. S. K. Goldsmith, T. C. Pellmar, A. M. Kleinman, and W. E. Bunney, *Reducing suicide: A national imperative* (Washington, DC: National Academy Press, 2002).
3. J. M. Harlow, "Recovery from the passage of an iron bar through the head," *History of Psychiatry* 4 (1993): 271–281.
4. A. Damasio, *Descartes' Error: Emotion, Reason, and the Human Brain* (New York: Putnam, 1994).
5. R. F. Baumeister, J. M. Twenge, and C. K. Nuss, "Effects of social exclusion on cognitive processes: Anticipated aloneness reduces intelligent thought," *Journal of Personality and Social Psychology* 83, no. 4 (2002): 817–827.
6. W. K. Campbell, E. A. Krusemark, K. A. Dyckman, A. B. Brunell, J. E. McDowell, J. M. Twenge, and B. A. Clementz, "A magnetoencephalography investigation of neural correlates for social exclusion and self-control," *Social Neuroscience* 1 (2006): 124–134.
7. R. F. Baumeister, C. N. DeWall, N. J. Ciarocco, and J. M. Twenge, "Social exclusion impairs self-regulation," *Journal of Personality and Social Psychology* 88 (2005): 589–604.
8. R. S. Weiss, *Loneliness: The experience of emotional and social isolation* (Cambridge, MA: MIT Press, 1973).
9. J. K. Maner, C. N. DeWall, R. F. Baumeister, and M. Schaller, "Does social exclusion motivate interpersonal reconnection? Resolving the 'porcupine problem,'" *Journal of Personality and Social Psychology* 92 (2007): 42–55.

10. J. M. Twenge, R. F. Baumeister, D. M. Tice, and T. S. Stucke, "If you can't join them, beat them: Effects of social exclusion on aggressive behavior," *Journal of Personality and Social Psychology* 81 (2001): 1058–69. K. Rotenberg, "Loneliness and interpersonal trust," *Journal of Social and Clinical Psychology* 13 (1994): 152–173.
11. J. M. Twenge, R. F. Baumeister, C. N. DeWall, N. J. Ciarocco, and J. M. Bartels, "Social exclusion decreases prosocial behavior," *Journal of Personality and Social Psychology* 92 (2007): 56–66. J. M. Twenge, K. R. Catanese, and R. F. Baumeister, "Social exclusion causes self-defeating behavior," *Journal of Personality and Social Psychology* 83 (2002): 606–615.
12. L. C. Hawkley and J. T. Cacioppo, "Aging and loneliness: Downhill quickly?" *Current Directions in Psychological Science* 16 (2007): 187–191.
13. Baumeister, DeWall, Ciarocco, and Twenge, "Social exclusion impairs self-regulation."
14. Weiss, *Loneliness.*
15. Ibid.
16. S. T. Boysen, G. G. Berntson, M. B. Hanna, and J. T. Cacioppo, "Quantity-based choices: Interference and symbolic representations in chimpanzees (Pan troglodytes)," *Journal of Experimental Psychology: Animal Behavior Processes* 22 (1996): 76–86.
17. J. Vitkus, and L. M. Horowitz, "Poor social performance of lonely people: Lacking a skill or adopting a role," *Journal of Personality and Social Psychology* 52 (1987): 1266–73.

CHAPTER FOUR: *Selfish Genes, Social Animals*

1. M. McPherson, L. Smith-Lovin, and M. T. Brashears, "Social isolation in America: Changes in core discussion networks over two decades," *American Sociological Review* 71 (2006): 353–375.
2. F. Hobbs and N. Stoops, *Demographic trends in the 20th century*, U.S. Census Bureau, Census 2000 Special Reports, Series CENSR-4 (Washington, DC: U.S. Government Printing Office, 2002).
3. Thomas Hobbes, *Leviathan*, Everyman ed. (1651; New York: Dutton, 1975), introduction by K. R. Minogue.
4. Ibid.
5. Charles Darwin, *Autobiography* (1887), in F. Darwin, ed., *The life and letters of Charles Darwin* (Whitefish, MT: Kessinger, 2004).
6. G. Williams, *Adaptation and Natural Selection* (Princeton: Princeton University Press, 1966).
7. R. F. Baumeister and C. N. DeWall, "The inner dimensions of social

exclusion: Intelligent thought and self-regulation among rejected persons," in K. D. Williams, J. P. Forgas, and W. von Hippel, eds., *The social outcast: Ostracism, social exclusion, rejection, and bullying* (New York: Psychology Press, 2005), 53–73.

8. The languages of the Kalahari rely on "click" sounds, as well as guttural sounds produced deep within the throat. Thus "!Kung" is pronounced "Gung," as if you were imitating water glugging through a drainpipe, but with a strong initial clicking sound on the hard "g."
9. Bruce Bowere, "Murder in good company," *Science News,* February 6, 1988.
10. M. Nowak, "Five rules for the evolution of cooperation," *Science* 314 (2006): 1560–63.
11. R. I. M. Dunbar and Suzanne Shultz, "Evolution and the social brain," *Science* 317 (September 7, 2007): 1344–47.
12. D. L. Cheney and R. M. Seyfarth, *Baboon metaphysics* (Chicago: University of Chicago Press, 2007).
13. Williams, *Adaptation and Natural Selection.*
14. R. L. Trivers, "Parental investment and sexual selection," in B. Campbell, ed., *Sexual selection and the descent of man, 1871–1971* (Chicago: Aldine, 1972), 136–179.
15. J. T. Cacioppo and L. C. Hawkley, "Loneliness," in M. R. Leary and R. H. Hoyle, eds., *Handbook of individual differences in social behavior* (New York: Guilford, in press); Dunbar and Shultz, "Evolution and the social brain."

CHAPTER FIVE: *The Universal and the Particular*

1. C. Tucker-Ladd, *Psychological self-help* (1996), retrieved June 19, 2007, from *www.psychologicalselfhelp.org.*
2. Marja Jylha, "Old Age and loneliness: Cross-sectional and longitudinal analyses in the Tampere Longitudinal Study on Aging," *Canadian Journal on Aging* 23, no. 2 (2004): 157–158.
3. M. B. Brewer and W. Gardner, "Who is this 'we'? Levels of collective identity and self representations," *Journal of Personality and Social Psychology* 71 (1996): 83–93.
4. Ibid.
5. L. C. Hawkley, M. W. Browne, and J. T. Cacioppo, "How can I connect with thee? Let me count the ways," *Psychological Science* 16 (2005): 798–804.
6. W. Mischel, Y. Shoda, and R. E. Smith, *Introduction to personality: Toward an integration,* 7th ed. (New York: Wiley, 2004).
7. Clinical depression is a complex diagnosis in which the individual must

exhibit a variety of specific symptoms such as difficulty making decisions, difficulty sleeping, or loss of appetite. Depressed affect is a more intuitive, commonsense designation that includes simply feeling down, however briefly.

8. C. Segrin, "Interpersonal communication problems associated with depression and loneliness," in P. A. Andersen and L. K. Guerrero, eds., *Handbook of communication and emotion: Research, theory, applications, and contexts* (San Diego: Academic Press, 1998), 215–242.
9. L. S. Radloff, "The CES-D Scale: A self-report depression scale for research in the general population," *Applied Psychological Measurement* 1 (1977): 385–401.
10. R. S. Weiss, ed., *Loneliness: The experience of emotional and social isolation* (Cambridge, MA: MIT Press, 1973); J. T. Cacioppo, L. C. Hawkley, J. M. Ernst, M. Burleson, G. G. Berntson, B. Nouriani, and D. Spiegel, "Loneliness within a nomological net: An evolutionary perspective," *Journal of Research in Personality* 40 (2006): 1054–85.
11. P. Watson and P. Andrews, "Toward a revised evolutionary adaptationist analysis of depression: The social navigation hypothesis," *Journal of Affective Disorders* 72 (2002): 1–14.
12. G. L. Engel, "The clinical application of the biopsychosocial model," *American Journal of Psychiatry* 137 (1980): 535–544.
13. J. S. Price, L. Sloman, R. Gardner, P. Gilbert, and P. Rhode, "The social competition hypothesis of depression," *British Journal of Psychiatry* 164 (1994): 309–315.
14. E. H. Hagan, "The function of postpartum depression," *Evolution and Human Behavior* 20 (1999): 325–359.
15. N. B. Allen and P. B. T. Badcock, "The social risk hypothesis of depressed mood: Evolutionary, psychosocial, and neurobiological perspectives," *Psychological Bulletin* 129 (2003): 887–913.
16. J. T. Cacioppo, J. M. Ernst, M. H. Burleson, M. K. McClintock, W. B. Malarkey, L. C. Hawkley, R. B. Kowalewski, A. Paulsen, J. A. Hobson, K. Hugdahl, D. Spiegel, and G. G. Berntson, "Lonely traits and concomitant physiological processes: The MacArthur Social Neuroscience Studies," *International Journal of Psychophysiology* 35 (2000): 143–154.
17. J. M. Ernst and J. T. Cacioppo, "Lonely hearts: Psychological perspectives on loneliness," *Applied and Preventive Psychology 8* (1998): 1–22; M. R. Leary and R. F. Baumeister, "The nature and function of self-esteem: Sociometer theory," in M. P. Zanna, ed., *Advances in experimental social psychology*, vol. 32 (San Diego: Academic Press, 2000), 1–62; M. R. Leary, E. S. Tambor, S. K. Terdal, and D. L. Downs, "Self-esteem as an interpersonal monitor," *Journal of Personality and Social Psychology* 68 (1995): 518–530.
18. Cacioppo et al., "Loneliness within a nomological net."

19. S. M. Kosslyn, W. L. Thompson, M. F. Costantini-Ferrando, N. M. Alpert, and D. Spiegel, "Hypnotic visual illusion alters color processing in the brain," *American Journal of Psychiatry* 157 (2000): 1279–84.
20. J. T. Cacioppo, M. E. Hughes, L. J. Waite, L. C. Hawkley, and R. A. Thisted, "Loneliness as a specific risk factor for depressive symptoms: Cross sectional and longitudinal analyses," *Psychology and Aging* 21 (2006): 140–151.
21. Ibid.

CHAPTER SIX: *The Wear and Tear of Loneliness*

1. R. Lewontin, *The triple helix: Gene, organism, and environment* (Cambridge, MA: Harvard University Press, 2002). J. Irving, *The world according to Garp* (1978; New York: Ballantine, 1990), 618.
2. T. C. Pellmar, E. N. Brandt, and M. A. Baird, "Health and behavior: The interplay of biological, behavioral, and social influences: Summary of an Institute of Medicine report," *American Journal of Health Promotion* 16, no. 4 (2001): 206–219.
3. L. F. Berkman and S. L. Syme, "Social networks, host resistance and mortality: A nine-year follow-up study of Alameda County residents," *American Journal of Epidemiology* 109, no. 2 (1979): 186–204.
4. J. S. House, K. R. Landis, and D. Umbertson, "Social relationships and health," *Science* 241 (1988): 540–545.
5. D. Russell, E. Cutrona, A. De La Mora, and R. B. Wallace, "Loneliness and nursing home admission among rural older adults," *Psychology and Aging* 12 (1997): 574–589.
6. L. Wheeler, H. Reis, and J. B. Nezlek, "Loneliness, social interaction, and sex roles," *Journal of Personality and Social Psychology* 45 (1983): 943–953. L. C. Hawkley, M. H. Burleson, G. G. Berntson, and J. T. Cacioppo, "Loneliness in everyday life: Cardiovascular activity, psychosocial context, and health behaviors," *Journal of Personality and Social Psychology* 85 (2003): 105–120.
7. J. T. Cacioppo, L. C. Hawkley, G. G. Berntson, J. M. Ernst, A. C. Gibbs, R. Stickgold, and J. A. Hobson, "Lonely days invade the nights: Social modulation of sleep efficiency," *Psychological Science* 13 (2002): 384–387. J. T. Cacioppo, L. C. Hawkley, L. E. Crawford, J. M. Ernst, M. H. Burleson, R. B. Kowalewski, W. B. Malarkey, E. Van Cauter, and G. G. Berntson, "Loneliness and health: Potential mechanisms," *Psychosomatic Medicine 64* (2002): 407–417.
8. P. A. Nakonezny, R. B. Kowalewski, J. M. Ernst, L. C. Hawkley, D. L. Lozano, D. A. Litvack, G. G. Berntson, J. J. Sollers III, P. Kizakevich, J. T. Cacioppo, and W. R. Lovallo, "New ambulatory impedance cardiograph

validated against the Minnesota impedance cardiograph," *Psychophysiology* 38 (2001): 465–474. The monitor was developed for us by a team from the MacArthur Network led by Bill Lovallo.

9. R. W. Frenck Jr., E. H. Blackburn, and K. M. Shannon, "The rate of telomere sequence loss in human leukocytes varies with age," *Proceedings of the National Academy of Sciences* 95 (1998): 5607–10.
10. M. Marmot, *The Status Syndrome* (New York: Times Books, 2004).
11. M. H. Hecker, M. A. Chesney, G. W. Black, and N. Frautschi, "Coronary-prone behaviors in the Western Collaborative Group Study," *Psychosomatic Medicine* 50 (1988): 153–164.
12. J. M. Ernst and J. T. Cacioppo, "Lonely hearts: Psychological perspectives on loneliness," *Applied and Preventive Psychology* 8 (1998): 1–22.
13. M. D. Boltwood, C. B. Taylor, M. B. Burke, H. Grogin, and J. Giacomini, "Anger report predicts coronary artery vasomotor response to mental stress in atherosclerotic segments," *American Journal of Cardiology* 72 (1993): 1361–65.
14. G. Ironson, C. B. Taylor, M. Boltwood, T. Bartzokis, C. Dennis, M. Chesney, S. Spitzer, and G. M. Segall, "Effects of anger on left ventricular ejection fraction in coronary artery disease," *American Journal of Cardiology* 70 (1992): 281–285.
15. N. A. Christakis and J. H. Fowler, "The spread of obesity in a large social network over 32 years," *New England Journal of Medicine* 357, no. 4 (July 26, 2007): 370–379.
16. N. E. Adler, M. A. Chesney, C. S. Folkman, R. L. Kahn, and S. L. Syme, "Socioeconomic status and health," *American Psychologist* 49, no. 1 (1994): 15–24. G. A. Kaplan and J. E. Keil, "Socioeconomic factors and cardiovascular disease: A review of the literature," *Circulation* 88 (1993): 141–142.
17. W. B. Cannon, "The role of emotions in disease," *Annals of Internal Medicine* 11 (1936): 1453–65.
18. G. G. Berntson and J. T. Cacioppo, "From homeostasis to allodynamic regulation," in J. T. Cacioppo, L. G. Tassinary, and G. G. Berntson, eds., *Handbook of psychophysiology*, 2nd ed. (Cambridge: Cambridge University Press, 2000), 459–481. P. Sterling and J. Eyer, "Allostasis: A new paradigm to explain arousal pathology," in S. Fisher and J. Reason, eds., *Handbook of life stress, cognition and health* (New York: Wiley, 1988), 629–649.
19. T. E. Seeman, B. S. McEwen, J. W. Rowe, and B. H. Singer, "Allostatic load as a marker of cumulative biological risk: MacArthur studies of successful aging," *Proceedings of the National Academy of Sciences* 98 (1997): 4770–75. B. S. McEwen, "Protective and damaging effects of stress mediators," *New England Journal of Medicine* 338 (1998): 171–179.
20. J. T. Cacioppo, M. E. Hughes, L. J. Waite, L. C. Hawkley, and R. A. Thisted, "Loneliness as a specific risk factor for depressive symptoms: Cross-sectional and longitudinal analyses," *Psychology and Aging* 21 (2006):

140–151. Loneliness predicts hypertension and cardiovascular disease in the English Longitudinal Study of Ageing; J. Smith, personal communication, October 2007. L. C. Hawkley, C. M. Masi, J. D. Berry, and J. T. Cacioppo, "Loneliness is a unique predictor of age-related differences in systolic blood pressure," *Psychology and Aging* 21 (2006): 152–164.

21. L. C. Hawkley and J. T. Cacioppo, "Aging and loneliness: Downhill quickly?" *Current Directions in Psychological Science* 16 (2007): 187–191.
22. Ibid.
23. Hawkley, Burleson, Berntson, and Cacioppo, "Loneliness in everyday life." J. T. Cacioppo, J. M. Ernst, M. H. Burleson, M. K. McClintock, W. B. Malarkey, L. C. Hawkley, R. B. Kowalewski, A. Paulsen, J. A. Hobson, K. Hugdahl, D. Spiegel, and G. G. Berntson, "Lonely traits and concomitant physiological processes: The MacArthur social neuroscience studies," *International Journal of Psychophysiology* 35 (2000): 143–154.
24. Hawkley and Cacioppo, "Aging and loneliness: Downhill quickly?"
25. P. L. Schnall, P. A. Landsbergis, and D. Baker, "Job strain and cardiovascular disease," *Annual Review of Public Health* 15 (1994): 381–411.
26. Cacioppo et al., "Lonely traits and concomitant physiological processes." Hawkley et al., "Loneliness in everyday life."
27. Cacioppo et al., "Lonely traits and concomitant physiological processes."
28. Ibid.
29. J. T. Cacioppo and G. G. Berntson, "A bridge linking social psychology and the neurosciences," in Paul A. M. Van Lange, ed., *Bridging social psychology: The benefits of transdisciplinary approaches* (Hillsdale, NJ: Erlbaum, 2006).
30. Hawkley et al., "Loneliness is a unique predictor of age-related differences in systolic blood pressure."
31. R. Glaser, J. K. Kiecolt-Glaser, C. E. Speicher, and J. E. Holliday, "Stress, loneliness, and changes in herpes virus latency," *Journal of Behavioral Medicine* 8, no. 3 (September 1985): 249–260. S. D. Pressman, S. Cohen, G. E. Miller, A. Barkin, B. S. Rabin, and J. J. Treanor, "Loneliness, social network size, and immune response to influenza vaccination in college freshmen," *Health Psychology* 24 (2005): 297–306.
32. L. C. Hawkley, J. A. Bosch, C. G. Engeland, P. T. Marucha, and J. T. Cacioppo, "Loneliness, dysphoria, stress and immunity: A role for cytokines," in N. P. Plotnikoff, R. E. Faith, and A. J. Murgo, eds., *Cytokines: Stress and immunity*, 2nd ed. (Boca Raton, FL: CRC Press, 2007), 67–86.
33. E. K. Adam, L. C. Hawkley, B. M. Kudielka, and J. T. Cacioppo, "Day-to-day dynamics of experience: Cortisol associations in a population-based sample of older adults," *Proceedings of the National Academy of Sciences* 103 (2006): 17058–63. S. W. Cole, L. C. Hawkley, J. M. Arevalo, C. Y. Sung, R. M. Rose, and J. T. Cacioppo, "Social regulation of gene expression in

human leukocytes," *Genome Biology* 8, no. 9 (2007): R189. Hawkley et al., "Loneliness, dysphoria, stress, and immunity."

34. A. Sherwood, C. A., Dolan, and K. C. Light, "Hemodynamics of blood pressure responses during active and passive coping," *Psychophysiology* 27 (1990): 656–668.
35. Cacioppo et al., "Loneliness and health: Potential mechanisms." Hawkley et al., "Loneliness in everyday life."
36. Hawkley et al., "Loneliness is a unique predictor of age-related differences in systolic blood pressure." Cacioppo et al., "Loneliness and health." Jim Smith, personal communication, 2007.
37. Lewontin, *Triple Helix*, 104.
38. K. Spiegel, R. Leprout, and E. Van Cauter, "Impact of sleep debt on metabolic function," *Lancet* 354 (1999): 1435–39.
39. Cacioppo et al., "Lonely days invade the nights."
40. Hawkley and Cacioppo, "Aging and loneliness: Downhill quickly?"

CHAPTER SEVEN: *Sympathetic Threads*

1. Henry Melvill, *Best thoughts of best thinkers*, Penny Pulpit Sermons no. 2,365 (Cleveland, 1904).
2. Nicholas A. Christakis and James H. Fosler, "The spread of obesity in a large social network over 32 years," *New England Journal of Medicine* 357 (July 26, 2007): 370–379.
3. D. P. Phillips, T. E. Ruth, and L. M. Wagner, "Psychology and survival," *Lancet* 342 (1993): 1142–45.
4. Antonio Damasio, *Descartes' error* (New York: Putnam, 1994); L. W. Barasalou, "Cognitive and neural contributions to understanding the conceptual system," *Current Directions in Psychological Science*, in press.
5. W. James, *The principles of psychology* (New York: Henry Holt, 1890).
6. C. E. Cornell, J. Rodin, and H. P. Weingarten, "Stimulus-induced eating when satiated," *Physiology and Behavior* 45 (1989): 695–704.
7. S. N. Haber and P. R. Barchas, "The regulatory effect of social rank on behavior after amphetamine administration," in P. R. Barchas, ed., *Social hierarchies: Essays toward a sociophysiological perspective* (Westport, CT: Greenwood, 1983), 119–132.
8. Robin Marantz Henig, "The real transformers," *New York Times Magazine*, July 29, 2007.
9. A. N. Meltzoff and M. K. Moore, "Imitation of facial and manual gestures by human neonates," *Science* 198 (1977): 75–78. M. Myowa-Yamakoski, M. Tomonaga, M. Tanaka, and T. Matsuzawa, "Imitation in neonatal chimpanzees," *Development Science* 7 (2004): 437–442.

10. Pier F. Ferrari, Elisabetta Visalberghi, Annika Paukner, Leonardo Fogassi, Angela Ruggiero, and Stephen J. Suomi, "Neonatal imitation in rhesus macaques," *PLOS Biology* 4, no. 9 (September 2006): e302.
11. L. B. Adamson and J. E. Frick, "The still face: A history of a shared experimental paradigm," *Applied Psychology and Management* 4, no. 4 (2003): 451–473.
12. E. Hatfield, J. T. Cacioppo, and R. L. Rapson, *Emotional contagion* (New York: Cambridge University Press, 1994), 240.
13. M. La France and M. Broadbent, "Group rapport: Posture sharing as a nonverbal indicator," *Group and Organization Studies* 1 (1976): 328–333.
14. F. J. Bernieri, "Coordinated movement and rapport in teacher-student interactions," *Journal of Nonverbal Behavior* 12 (1988): 120–138.
15. D. Byrne, *The attraction paradigm* (New York: Academic Press, 1971).
16. R. E. Maurer and J. H. Tindall, "Effect of postural congruence on client's perception of counselor empathy," *Journal of Counseling Psychology* 30 (1983): 158–163. J. L. Lakin and T. L. Chartrand, "Using nonconscious behavioral mimicry to create affiliation and rapport," *Psychological Science* 14 (2003): 334–339.
17. Lakin and Chartrand, "Using nonconscious behavioral mimicry to create affiliation and rapport."
18. M. R. Leary, C. A. Cottrell, and M. Phillips, "Deconfounding the effects of dominance and social acceptance on self-esteem," *Journal of Personality and Social Psychology* 81 (2001): 898–909. W. L. Gardner, "Social exclusion and selective memory: How the need to belong influences memory for social events," *Personality and Social Psychology Bulletin* 26 (2000): 486–496.
19. K. D. Williams, C. K. T. Cheung, and W. Choi, "Cyberostracism: Effects of being ignored over the Internet," *Journal of Personality and Social Psychology* 79 (2000): 748–762.
20. K. D. Williams and K. L. Sommer, "Social ostracism by coworkers: Does rejection lead to loafing or compensation?" *Personality and Social Psychology Bulletin* 23 (1997): 693–706.
21. S. E. Taylor, *The tending instinct: How nurturing is essential to who we are and how we live* (New York: Time Books, 2002).
22. N. S. Wingreen and S. A. Levin, "Cooperation among microorganisms," *PLOS Biology*, 4, no. 9 (2006): 299.
23. C. E. Taylor and M. T. McGuire, "Reciprocal altruism: Fifteen years later," *Ethology and Sociobiology* 9 (1988): 67–72.
24. F. de Waal, *Our inner ape* (New York: Riverhead, 2006).
25. C. Darwin, *The descent of man and selection in relation to sex* (1874; Chicago: Rand McNally, 1974), 613. A. Damasio, *Descartes' error* (New York: Putnam, 1994).

CHAPTER EIGHT: *An Indissociable Organism*

1. J. E. Swain, J. P. Lorberbaum, S. Kose, and L. Strathhearn, "Brain basis of early parent-infant interactions: Psychology, physiology, and in vivo functional neuroimaging studies," *Journal of Child Psychology and Psychiatry* 48, nos. 3/4 (2007): 262–287.
2. H. F. Harlow and R. Zimmerman, "Affectional responses in the infant monkey," *Science* 130 (1959): 421–432.
3. University of Wisconsin, *The Why Files.org 087/mother/4.html.*
4. K. Z. Lorenz, "Der Kumpan in der Umwelt des Vogels" (1935), English translation in *Instinctive behavior: The development of a modern concept*, trans. and ed. Claire H. Schiller (New York: International Universities Press, 1957).
5. M. D. S. Ainsworth, M. C. Blehar, E. Waters, and S. Wall, *Patterns of attachment: A psychological study of the strange situation* (Hillsdale, NJ: Erlbaum, 1978).
6. J. Kagan and N. Snidman, *The long shadow of temperament* (Cambridge, MA: Harvard University Press, 2004).
7. P. Ekman and R. J. Davidson, *The nature of emotion: Fundamental questions* (New York: Oxford University Press, 1994), xiv, 496.
8. R. J. Davidson, P. Ekman, C. Saron, J. Senulis, and W. V. Friesen, "Emotional expression and brain physiology I: Approach/withdrawal and cerebral asymmetry," *Journal of Personality and Social Psychology* 58 (1990): 330–341.
9. N. A. Fox, K. H. Rubin, S. D. Calkins, T. R. Marshall, R. J. Coplan, S. W. Porges, J. M. Long, and S. Stewart, "Frontal activation asymmetry and social competence at four years of age," *Child Development* 66 (1995): 1770–84.
10. A. J. Tomarken, R. J. Davidson, R. E. Wheeler, and L. Kinney, "Psychometric properties of resting anterior EEG asymmetry: Temporal stability and internal consistency," *Psychophysiology* 29 (1992): 576–592. See also S. K. Sutton and R. J. Davidson, "Prefrontal brain asymmetry: A biological substrate of the behavioral approach and inhibition systems," *Psychological Science* 8, no. 3 (1997): 204–210.
11. A. Damasio, *Descartes' Error* (New York: Putnam, 1994), xvi–xvii.
12. T. R. Insel and L. E. Shapiro, "Oxytocin receptor distribution reflects social organization in monogamous and polygamous voles," *Proceedings of the National Academy of Sciences* 89 (1992): 5981–85. T. R. Insel, Z. Wang, and C. F. Ferris, "Patterns of brain vasopressin receptor distribution associated with social organization in microtine rodents," *Journal of Neuroscience* 14 (1994): 5381–92. M. M. Lim, Z. Wang, D. E. Olazábal, X. Ren, E. F. Terwilliger, and L. J. Young, "Enhanced partner preference in

promiscuous species by manipulating the expression of a single gene," *Nature* 429 (2004): 754–757.

13. K. Uvnas-Moberg, *The oxytocin factor* (Cambridge, MA: Da Capo, 2003).
14. K. Uvnas-Moberg, "Oxytocin may mediate the benefits of positive social interaction and emotions," *Psychoneuroendocrinology* 23, no. 8 (1998): 819–835.
15. Uvnas-Moberg, *The oxytocin factor.*
16. Ibid.
17. Sam Roberts, "The shelf life of bliss," *New York Times*, July 1, 2007.
18. P. V. Bradford, *Ota Benga: The Pygmy in the zoo* (New York: St. Martin's, 1992).

CHAPTER NINE: *Knowing Thyself, among Others*

1. C. Darwin, *The expression of the emotions in men and animals* (1872; Chicago: University of Chicago Press, 1965).
2. F. de Waal, *Our inner ape* (New York: Riverhead, 2006).
3. N. K. Humphrey, *A history of the mind: Evolution and the birth of consciousness* (New York: Simon and Schuster, 1992).
4. A. Smith, *The theory of moral sentiments* (1752; Cambridge: Cambridge University Press, 2002), 12.
5. L. Carr, M. Iacoboni, M. C. Dubeau, J. C. Mazziotta, and G. L. Lenzi, "Neural mechanisms of empathy in humans: A relay from neural systems for imitation to limbic areas," *Proceedings of the National Academy of Sciences* 100 (2003): 5497–5502.
6. G. Di Pellegrino, L. Fadiga, L. Fogassi, V. Gallese, and G. Rizzolatti, "Understanding motor events: A neurophysiological study," *Experimental Brain Research* 91, no. 1 (1992): 176–180. S. Blakeslee, "Cells that read minds," *New York Times*, January 10, 2006.
7. V. Gallese, L. Fadiga, L. Fogassi, and G. Rizzolatti, "Action recognition in the premotor cortex," *Brain* 119, no. 2, 593–609, cited in L. Winerman, "The mind's mirror," *APA Online* 36, no. 9 (2005). M. A. Umilita, E. Kohler, V. Gallese, L. Fogassi, L. Fadiga, C. Keysers, and G. Rizzolatti, "I know what you are doing: A neurophysiological study," *Neuron* 31, no. 1 (2001): 155–165. G. R. Semin and J. T. Cacioppo, "Grounding social cognition: Synchronization, coordination, and co-regulation," in G. R. Semin and E. R. Smith, eds., *Embodied grounding: Social, cognitive, affective, and neuroscientific approaches* (New York: Cambridge University Press, in press).
8. G. Rizzolatti and L. Craighero, "The mirror-neuron system," *Annual Review of Neuroscience* 27 (2004): 169–192.
9. G. Buccino, F. Lui, N. Canessa, I. Patteri, G. Lagravinese, F. Benuzzi,

C. A. Porro, and G. Rizzolatti, "Neural circuits involved in the recognition of actions performed by nonconspecifics: An fMRI study," *Journal of Cognitive Neuroscience* 16 (2004): 114–126.

10. B. Wicker, C. Keysers, J. Plailly, J. P. Royet, V. Gallese, and G. Rizzolatti, "The common neural basis of seeing and feeling disgust," *Neuron* 40, no. 3 (2003): 655–664, reported in L. Winerman, "The mind's mirror," *APA Online* 36, no. 9 (2005).
11. G. Buccino, F. Binkofski, G. R. Fink, L. Fadiga, L. Fogassi, V. Gallese, R. J. Seitz, K. Zilles, G. Rizzolatti, and H. J. Freund, "Action observation activates premotor and parietal areas in a somatotopic manner: An fMRI study," *European Journal of Neuroscience* 13 (2001): 400–404.
12. E. Hatfield, J. T. Cacioppo, and R. L. Rapson, *Emotional contagion* (New York: Cambridge University Press, 1994).
13. J. T. Cacioppo and G. G. Berntson, *Social neuroscience* (New York: Psychology Press, 2005). H. Fukui, T. Murai, J. Shinozaki, T. Aso, H. Fukuyama, T. Hayashi, and T. Hanakawa, "The neural basis of social tactics: An fMRI study," *NeuroImage* 32 (2006): 913–920.
14. D. Tankersley, C. J. Stowe, and S. A. Huettel, "Altruism is associated with an increased neural response to agency," *Nature Neuroscience* 10 (2007): 150–151.
15. G. G. Berntson, A. Bechara, H. Damasio, D. Tranel, and J. T. Cacioppo, "Amygdala contribution to selective dimensions of emotion," *Social, Cognitive, and Affective Neuroscience* 2 (2007): 123–129.
16. K. Grill-Spector, N. Knouf, and N. Kanwisher, "The fusiform face area subserves face perception, not generic within-category identification," *Nature Neuroscience* 7, no. 5 (2004): 555–562. N. Kanwisher, J. McDermott, and M. M. Chun, "The fusiform face area: A module in human extrastriate cortex specialized for face perception," *Journal of Neuroscience* 17, no. 11 (1997): 4302–11.
17. M. L. Phillips et al., "A specific neural substrate for perceiving facial expressions of disgust," *Nature* 389, no. 6650 (1997): 495–498. J. Decety and C. Lamm, "The biological bases of empathy," in G. G. Berntson and J. T. Cacioppo, eds., *Handbook of neuroscience for the behavioral sciences* (New York: Wiley, in press). R. Adolphs, "Social cognition and the human brain," *Trends in Cognitive Sciences* 3, no. 12 (1999): 469–479. H. C. Breiter et al., "Response and habituation of the human amygdala during visual processing of facial expression," *Neuron* 17 (1996): 875–887.
18. J. S. Morris, A. Ohman, and R. J. Dolan, "Conscious and unconscious emotional learning in the human amygdala," *Nature 393* (1998): 467–470. P. J. Whalen et al., "Masked presentations of emotional facial expressions modulate amygdala activity without explicit knowledge," *Journal of Neuroscience* 18 (1998): 411–418.

19. C. J. Norris and J. T. Cacioppo, "I know how you feel: Social and emotional information processing in the brain," in E. Harmon-Jones and P. Winkielman, eds., *Social neuroscience* (New York: Guilford, 2007), 84–105.
20. C. J. Norris, E. E. Chen, D. C. Zhu, S. L. Small, and J. T. Cacioppo, "The interaction of social and emotional processes in the brain," *Journal of Cognitive Neuroscience* 16 (2004): 1818–29. J. C. Britton, K. L. Phan, S. F. Taylor, R. C. Welsh, K. C. Berridge, and I. Liberzon, "Neural correlates of social and nonsocial emotions: An fMRI study," *NeuroImage* 31 (2006): 397–409.
21. Norris and Cacioppo, "I know how you feel."
22. R. I. M. Dunbar and S. Shultz, "Evolution in the social brain," *Science* 317 (September 7, 2007): 1344–47.
23. D. M. Buss, *Handbook of evolutionary psychology* (New York: Wiley, 2005).
24. R. Adolphs and M. Spezio, "The neuroscience of social cognition," in Berntson and Cacioppo, eds., *Handbook of neuroscience for the behavioral sciences.*
25. Berntson et al., "Amygdala contribution to selective dimensions of emotion."
26. T. A. Ito and J. T. Cacioppo, "Electrophysiological evidence of implicit and explicit categorization processes," *Journal of Experimental Social Psychology* 36 (2000): 660–676.
27. H. L. Gallagher and C. D. Frith, "Functional imaging of theory of mind," *Trends in Cognitive Sciences* 7, no. 2 (2003): 77–83.
28. W. L. Gardner, C. L. Pickett, V. Jefferis, and M. Knowles, "On the outside looking in: Loneliness and social monitoring," *Personality and Social Psychology Bulletin* 31, no. 11 (2005): 1549–60.
29. C. L. Pickett and W. L. Gardner, "The social monitoring system: Enhanced sensitivity to social cues as an adaptive response to social exclusion," in K. D. Williams, J. P. Forgas, and W. von Hippel, eds., *The social outcast: Ostracism, social exclusion, rejection, and bullying* (New York: Psychology Press, 2005), 214–226.
30. Ibid.
31. J. T. Cacioppo, C. J. Norris, J. Decety, G. Monteleone, and H. C. Nusbaum, "In the eye of the beholder: Individual differences in loneliness predict neural responses to social stimuli," *Journal of Cognitive Neuroscience* (in press).
32. J. T. Cacioppo, J. M. Ernst, M. H. Burleson, M. K. McClintock, W. B. Malarkey, L. C. Hawkley, R. B. Kowalewski, A. Paulsen, J. A. Hobson, K. Hugdahl, D. Speigel, and G. G. Berntson, "Lonely traits and concomitant physiological processes: The MacArthur social neuroscience studies," *International Journal of Psychophysiology* 35 (2000): 143–154.
33. S. L. Gable, G. Gonzaga, and A. Strachman, "Will you be there for me when things go right? Social support for positive events," *Journal of Person-*

ality and Social Psychology 91 (2006): 904–917. M. D. Johnson et al., "Problem-solving skills and affective expressions as predictors of change in marital satisfaction," *Journal of Consulting and Clinical Psychology* 73, no. 1 (2005): 15–27.

34. P. L. Jackson, A. N. Meltzoff, and J. Decety, "How do we perceive the pain of others? A window into the neural processes involved in empathy," *NeuroImage* 24 (2005): 771–779.
35. P. L. Jackson and J. Decety, "Motor cognition: A new paradigm to study self-other interactions," *Current Opinion in Neurobiology* 14, no. 2 (2004): 259–263.
36. J. Decety and C. Lamm, "The biological bases of empathy," in G. G. Berntson and J. T. Cacioppo, eds., *Handbook of neuroscience for the behavioral sciences* (New York: John Wiley & Sons, in press).
37. D. Schiller, ed., *The little Zen companion* (New York: Workman, 1994).

CHAPTER TEN: *Conflicted by Nature*

1. C. N. Macrae, J. Moran, T. Heatherton, J. Banfield, and W. Kelley, "Medial prefrontal activity predicts memory for self," *Cerebral Cortex* 14 (2004): 647–654. K. N. Ochsner, K. Knierim, D. Ludlow, J. Hanelin, T. Ramachandran, and S. Mackey, "Reflecting upon feelings: An fMRI study of neural systems supporting the attribution of emotion to self and other," *Journal of Cognitive Neuroscience* 16, no. 10 (2004): 1746–72.
2. H. Shintel, J. T. Cacioppo, and H. Nusbaum, "Accentuate the negative, eliminate the positive? Individual differences in attentional bias to positive and negative information," presented at the 47th Annual Meeting of the Psychonomic Society, Houston, TX, November 2006.
3. J. Kruger and T. Gilovich, "'Naïve cynicism' in everyday theories of responsibility assessment: On biased assumptions of bias," *Journal of Personality and Social Psychology* 76 (1999): 743–753. L. Ross, D. Greene, and P. House, "The false consensus effect: An egocentric bias in social perception and attributional processes," *Journal of Experimental Social Psychology* 13 (1977): 279–301. W. J. McGuire, "The probabilogical model of cognitive structure and attitude change," in R. E. Petty, T. M. Ostrom, and T. C. Brock, eds., *Cognitive responses in persuasion* (Hillsdale, NJ: Erlbaum, 1981), 291–307.
4. M. Ross and F. Sicoly, "Egocentric biases in availability and attribution," *Journal of Personality and Social Psychology* 37 (1979): 322–336. E. Vaughan, "Chronic exposure to an environmental hazard: Risk perceptions and self-protective behavior," *Health Psychology* 3 (1992): 431–457. E. F. Loftus, *Eyewitness testimony* (Cambridge, MA: Harvard University Press, 1996).
5. C. A. Anderson, R. S. Miller, A. L. Riger, J. C. Dill, and C. Sedikides,

"Behavioral and characterological attributional styles as predictors of depression and loneliness: Review, refinement, and test," *Journal of Personality and Social Psychology* 66 (1994): 549–558.

6. S. E. Taylor, J. S. Lerner, D. K. Sherman, R. M. Sage, and N. K. McDowell, "Portrait of the self-enhancer: Well-adjusted and well-liked or maladjusted and friendless?" *Journal of Personality and Social Psychology* 84 (2003): 165–176.
7. L. C. Hawkley, C. M. Masi, J. D. Berry, and J. T. Cacioppo, "Loneliness is a unique predictor of age-related differences in systolic blood pressure," *Psychology and Aging* 21 (2006): 152–164.
8. J. E. Nurmi and K. Salmela-Aro, "Social strategies and loneliness: A prospective study," *Personality and Individual Differences* 23, no. 2 (1997): 205–211. D. Damsteegt, "Loneliness, social provisions and attitude," *College Student Journal* 26, no. 1 (1992): 135–139. C. S. Crandall and C. Cohen, "The personality of the stigmatizer: Cultural world view, conventionalism, and self-esteem," *Journal of Research in Personality* 28 (1994): 461–480.
9. K. Rotenberg, "Loneliness and interpersonal trust," *Journal of Social and Clinical Psychology* 13 (1994): 152–173.
10. H. S. Sullivan, *The interpersonal theory of psychiatry* (New York: Norton, 1953), 261, quoted in R. S. Weiss, ed., *Loneliness: The experience of emotional and social isolation* (Cambridge, MA: MIT Press, 1973), 147.
11. J. Milton, *Paradise Lost* (1667), in M. Y. Hughes, ed., *John Milton: Complete Poems and Major Prose* (Indianapolis: Bobbs-Merrill, 1957), 217. W. Shakespeare, *Hamlet* (1603), act 2, scene 2.
12. N. Epley, A. Waytz, and J. T. Cacioppo, "On seeing human: A three-factor theory of anthropomorphism," *Psychological Review* 114 (2007): 864–886.
13. S. L. Murray and J. G. Holmes, "The (mental) ties that bind: Cognitive structures that predict relationship resilience," *Journal of Personality and Social Psychology* 77 (1999): 1228–44.
14. C. H. Solano, "Loneliness and perceptions of control: General traits versus specific attributions," *Journal of Social Behavior and Personality* 2, no. 2 (1987): 201–214.
15. S. Lau and G. E. Gruen, "The social stigma of loneliness: Effect of target person's and perceiver's sex," *Personality and Social Psychology Bulletin* 18 (1992): 182–189. K. J. Rotenberg and J. Kmill, "Perception of lonely and non-lonely persons as a function of individual differences in loneliness," *Journal of Social and Personal Relationships* 9 (1992): 325–330. K. J. Rotenberg, J. A. Gruman, and M. Ariganello, "Behavioral confirmation of the loneliness stereotype," *Basic and Applied Social Psychology* 24 (2002): 81–89.
16. S. L. Murray, G. M. Bellavia, P. Rose, and D. W. Griffin, "Once hurt, twice

hurtful: How perceived regard regulates daily marital interactions," *Journal of Personality and Social Psychology* 84 (2003): 126–147.

17. Rotenberg and Kmill, "Perception of lonely and non-lonely persons as a function of individual differences in loneliness."
18. M. T. Wittenberg and H. T. Reis, "Loneliness, social skills, and social perception," *Personality and Social Psychology Bulletin* 12, no. 1 (1986): 121–130. J. T. Cacioppo and L. C. Hawkley, "People thinking about people: The vicious cycle of being a social outcast in one's own mind," in K. D. Williams, J. P. Forgas, and W. von Hippel, eds., *The social outcast: Ostracism, social exclusion, rejection, and bullying* (New York: Psychology Press, 2005), 91–108.
19. S. Duck, K. Pond, and G. Leatham, "Loneliness and the evaluation of relational events," *Journal of Social and Personal Relationships* 11 (1994): 253–276.
20. C. M. Anderson and M. M. Martin, "The effects of communication motives, interaction involvement, and loneliness on satisfaction," *Small Group Research* 26, no. 1 (1995): 118–137.
21. A. Burt and R. Trivers, *Genes in conflict* (Cambridge, MA: Harvard University Press, 2006).

CHAPTER ELEVEN: *Conflicts in Nature*

1. R. L. Trivers, "Parent-offspring conflict," *American Zoologist* 14 (1974): 249–264, 261.
2. F. de Waal, *Our inner ape* (New York: Riverhead, 2006).
3. M. Doebeli, C. Hauert, and T. Killingback, "The evolutionary origin of cooperators and defectors," *Science* 306 (2004): 859–862.
4. O. Gurerk, B. Irlenbursch, and B. Rockenbach, "The competitive advantage of sanctioning institutions," *Science* 312 (April 7, 2006): 108–111.
5. Richard Dawkins, *The selfish gene* (New York: Oxford University Press, 1976).
6. B. Sinervo, A. Chaine, J. Clobert, R. Calsbeek, L. Hazard, L. Lancaster, A. G. McAdam, S. Alonzo, G. Corrigan, and M. E. Hochberg, "Self-recognition, color signals, and cycles of greenbeard mutualism and altruism," *Proceedings of the National Academy of Sciences* 103, no. 19 (2006): 7372–77.
7. W. Grossman, "New tack wins prisoner's dilemma," retrieved June 20, 2007, from *www.wired.com/culture/lifestyle/news/2004/10/65317.*
8. L. Cosimides and J. Tooby, "Evolutionary psychology and the generation of culture (part two)," *Ecology and Sociobiology* 10 (1989): 51–97.
9. M. A. Nowak, "Five rules for the evolution of cooperation," *Science* 314 (2006): 1560–63.

10. B. Carey, "Study links punishment to an ability to profit," *New York Times*, April 7, 2006.
11. E. Fehr and S. Gächter, "Altruistic punishment in humans," *Nature* 415 (2002): 137–140.
12. D. De Quervain, U. Fischbacher, V. Treyer, M. Schellhammer, U. Schnyder, A. Buck, and E. Fehr, "The neural basis of altruistic punishment," *Science* 305 (2004): 1254–58.
13. R. Trivers, "The evolution of reciprocal altruism," *Quarterly Review of Biology* 46, no. 1 (1971): 35–57, 49.
14. M. Wilson and M. Daly, "The age-crime relationship and the false dichotomy of biological versus sociological explanations," paper presented at a meeting of the Human Behavior and Evolution Society, Los Angeles, 1990.

CHAPTER TWELVE: *Three Adaptations*

1. K. E. Reed, "Early hominid evolution and ecological change through the African Plio-Pleistocene," *Journal of Human Evolution* 32 (1997): 289–322. S. Begley, "Beyond stones and bones," *Newsweek*, March 19, 2007, 52–58.
2. B. Heinrich, *Racing the antelope: What animals can teach us about running and life* (New York: Ecco, 2001).
3. I. Parker, "Swingers," *New Yorker*, July 30, 2007, 48–61.
4. M. A. Nowak, "Five rules for the evolution of cooperation," *Science* 314 (2006): 1560–63.
5. J. Silk, "Who are the more helpful, humans or chimpanzees?" *Science* 311 (2006): 1248–49.
6. Ibid.
7. E. Pennisi, "Social animals prove their smarts," *Science* 312, no. 5781 (2006): 1734–38.
8. Richard Dawkins, *The selfish gene* (New York: Oxford University Press, 1976); E. O. Wilson and C. Lumsden, *Genes, mind, and culture: The coevolutionary process* (Cambridge, MA: Harvard University Press, 1981).
9. F. de Waal, *Our inner ape* (New York: Riverhead, 2006).
10. Ibid., 54.
11. J. Goodall, *The Chimpanzees of Gombe: Patterns of Behavior* (Cambridge, MA: Belknap, 1986), cited in de Waal, *Our inner ape*.
12. de Waal, *Our inner ape*, 158.
13. A. Stravynski and R. Boyer, "Loneliness in relation to suicide ideation and parasuicide: A population-wide study," *Suicide and Life-Threatening Behavior* 31 (2001): 32–40; A. R. Rich and R. L. Bonner, "Concurrent validity of a stress-vulnerability model of suicidal ideation and behavior: A follow-up study," *Suicide and Life-Threatening Behavior* 17 (1987): 265–270.

14. J. M. Twenge, R. F. Baumeister, D. M. Tice, and T. S. Stucke, "Social exclusion causes self-defeating behavior," *Journal of Personality and Social Psychology* 83 (2001): 606–615. J. M. Twenge, K. R. Catanese, and R. F. Baumeister, "If you can't join them, beat them: Effects of social exclusion on aggressive behavior," *Journal of Personality and Social Psychology* 81 (2002): 1058–69. J. M. Twenge, K. R. Catanese, and R. F. Baumeister, "Social exclusion and the deconstructed state: Time perception, meaninglessness, lethargy, lack of emotion, and self-awareness," *Journal of Personality and Social Psychology* 85 (2003): 409–423.
15. J. T. Cacioppo, W. L. Gardner, and G. G. Berntson, "The affect system has parallel and integrative processing components: Form follows function," *Journal of Personality and Social Psychology* 76 (1999): 839–855.
16. E. Suh, E. Diener, and F. Fujita, "Events and subjective well-being: Only recent events matter," *Journal of Personality and Social Psychology* 70 (1996): 1091–1102.
17. P. Brickman, D. Coates, and R. Janoff-Bulman, "Lottery winners and accident victims: Is happiness relative?" *Journal of Personality and Social Psychology* 36 (1978): 917–927.
18. J. T. Cacioppo, L. C. Hawkley, A. Kalil, M. E. Hughes, L. Waite, and R. A. Thisted, "Happiness and the invisible threads of social connection: The Chicago Health, Aging, and Social Relations Study," in M. Eid and R. Larsen, eds., *The science of well-being* (New York: Guilford, 2008), 195–219.
19. Ibid.
20. J. T. Cacioppo, G. G. Berntson, A. Bechara, D. Tranel, and L. C. Hawkley, "Could an aging brain contribute to subjective well-being? The value added by a social neuroscience perspective," in A. Tadorov, S. T. Fiske, and D. Prentice, eds., *Social neuroscience: Toward understanding the underpinnings of the social mind* (New York: Oxford University Press, in press).
21. L. L. Carstensen, D. M. Isaacowitz, and S. T. Charles, "Taking time seriously: A theory of socioemotional selectivity," *American Psychologist* 54 (1999): 165–181.
22. A. M. Isen, "Positive affect and decision making," in M. Lewis and J. M. Haviland-Jones, eds., *Handbook of emotions*, 2nd ed. (New York: Guilford, 2000), 417–435.

CHAPTER THIRTEEN: *Getting It Right*

1. A. M. Isen, "Positive affect, cognitive processes, and social behavior," *Advances in Experimental Social Psychology*, 20 (1987): 203–253.
2. J. Masters, *The road past Mandalay: A personal narrative* (New York: Harper, 1961).

3. S. Sassoon, *Memoirs of an infantry officer* (London: Faber and Faber, 1930).
4. M. Wei, D. W. Russell, and R. A. Aakalik, "Adult attachment, social self-efficacy, self-disclosure, loneliness, and subsequent depression for freshman college students: A longitudinal study," *Journal of Counseling Psychology* 52 (2005): 602–614. J. T. Cacioppo, M. E. Hughes, L. J. Waite, L. C. Hawkley, and R. A. Thisted, "Loneliness as a specific risk factor for depressive symptoms: Cross sectional and longitudinal analyses," *Psychology and Aging* 21 (2006): 140–151.
5. I. L. Martinez, K. Frick, T. A. Glass, M. Carlson, E. Tanner, M. Ricks, and L. Fried, "Engaging older adults in high-impact volunteering that enhances health: Recruitment and retention in the Experience Corps Baltimore," *Journal of Urban Health* 83, no. 5 (2006): 941–953.
6. R. Niebuhr, *Moral man and immoral society* (New York: Scribner, 1932).
7. W. B. Swann Jr., K. L. McClarty, and P. J. Rentfrow, "Shelter from the storm? Flawed reactions to stress in precarious couples," *Journal of Social and Personal Relationships* (2007): 793–808.
8. S. L. Murray, G. Bellavia, P. Rose, and D. Griffin, "Once hurt, twice hurtful: How perceived regard regulates daily marital interaction," *Journal of Personality and Social Psychology* 84 (2003): 126–147.
9. J. M. Martz, J. Verette, X. B. Arriaga, L. F. Slovik, C. L. Cox, and C. E. Rusbult, "Positive illusion in close relationships," *Personal Relationships* 5 (1998): 159–181.
10. P. J. E. Miller, S. Niehuis, and T. L. Huston, "Positive illusions in marital relationships: A 13-year study," *Personality and Social Psychology Bulletin* 32, no. 12 (2006): 1579–94.
11. S. L. Gable, H. T. Reis, E. Impett, and E. R. Asher, "What do you do when things go right? The intrapersonal and interpersonal benefits of sharing positive events," *Journal of Personality and Social Psychology* 87 (2004): 228–245.

CHAPTER FOURTEEN: *The Power of Social Connection*

1. M. McPherson, L. Smith-Lovin, and M. T. Brashears, "Social isolation in America: Changes in core discussion networks over two decades," *American Sociological Review* 71 (2006): 353–375.
2. WHO World Mental Health Survey Consortium, "Prevalence, severity, and unmet need for treatment of mental disorders in the World Health Organization World Mental Health Surveys," *Journal of the American Medical Association* 291 (2004): 2581–90.
3. UNICEF Innocenti Research Centre, Florence, *An overview of child well being in rich countries*, United Nations Children's Fund, February 13, 2007.

4. L. Margulis and D. Sagan, *What is life?* (New York: Simon and Schuster, 1995).
5. R. S. Weiss, ed., *Loneliness: The experience of emotional and social isolation* (Cambridge, MA: MIT Press, 1973).
6. Ibid.
7. R. Putnam, *Bowling alone: The collapse and revival of American community* (New York: Simon and Schuster, 2000). J. Berger, "Homes too rich for firefighters who save them," *New York Times*, April 9, 2006.
8. S. Coontz, "Too close for comfort," *New York Times*, November 7, 2006.
9. K. Vonnegut, *Slapstick* (New York: Delacorte, 1976).
10. R. Schuller, *My journey: From an Iowa farm to a cathedral of dreams* (San Francisco: Harper, 2002).
11. Samuel G. Freedman, "An unlikely megachurch lesson," *New York Times*, November 3, 2007.
12. J. Mahler, "The soul of the new exurb," *New York Times Magazine*, March 27, 2005.
13. J. Yardley, "China's path to modernity, mirrored in a troubled river," *New York Times*, November 19, 2006. Lippmann quoted in Putnam, *Bowling alone*, 379.
14. Nadime Kam, "Rosie," *Honolulu Star Bulletin*, October 13, 2000.
15. Ibid.
16. A. Rokach and H. Brock, "Coping with loneliness," *Journal of Psychology: Interdisciplinary and Applied* 192 (1998): 107–127. B. S. Cain, "Divorce among elderly women: A growing social phenomenon," *Social Casework—Journal of Contemporary Social Work* 69 (1988): 563–568. S. T. Michael, M. R. Crowther, B. Schmid, and R. S. Allen, "Widowhood and spirituality: Coping responses to bereavement," *Journal of Women and Aging* 15 (2003): 145–165. P. Granqvist and B. Hagekkull, "Religiosity, adult attachment, and why 'singles' are more religious," *International Journal for the Psychology of Religion* 10 (2000): 111–123.
17. J. M. Siegel, "Stressful life events and use of physician services among the elderly: The moderating role of pet ownership," *Journal of Personality and Social Psychology* 58 (1990): 1081–86. J. M. Siegel, F. J. Angulo, R. Detels, J. Wesch, and A. Mullen, "AIDS diagnosis and depression in the multicenter AIDS cohort study: The ameliorating impact of pet ownership," *AIDS Care* 11 (1999): 157–170. K. Allen, J. Blascovich, and W. B. Mendes, "Cardiovascular reactivity and the presence of pets, friends and spouses: The truth about cats and dogs," *Psychosomatic Medicine* 64 (2002): 727–739.
18. K. Allen, J. Blascovich, and W. B. Mendes, "Cardiovascular reactivity and the presence of pets, friends and spouses: The truth about cats and dogs," *Psychosomatic Medicine* 64 (2002): 727–739.
19. N. Epley, S. Akalis, A. Waytz, and J. T. Cacioppo, "Creating social connec-

tion through inferential reproduction: Loneliness and perceived agency in gadgets, gods, and greyhounds," *Psychological Science* 19 (2008): 114–120. N. Epley, A. Waytz, S. Akalis, and J. T. Cacioppo, "When we need a human: Motivational determinants of anthropomorphism," *Social Cognition* 26 (2008): 143–155. N. Epley and S. Akalis, "Detecting versus enhancing anthropomorphic agents: The divergent effects of fear and loneliness," paper presented at a meeting of the Society for Personality and Social Psychology, New Orleans, February 2005.

20. S. L. Brown, R. M. Nesse, J. S. House, and R. L. Utz, "Religion and emotional compensation: Results from a prospective study of widowhood," *Personality and Social Psychology Bulletin* 30 (2004): 1165–74.
21. F. Castelli, F. Happé, U. Frith, and C. D. Frith, "Movement and mind: A functional imaging study of perception and interpretation of complex intentional movement patterns," *NeuroImage* 12 (2000): 314–325.
22. A. Birgegard and P. Granqvist, "The correspondence between attachment to parents and God: Three experiments using subliminal separation cues," *Personality and Social Psychology Bulletin* 30 (2004): 1122–35. T. Cole and L. Leets, "Attachment styles and intimate television viewing: Insecurely forming relationships in a parasocial way," *Journal of Social and Personal Relationships* 16 (1999): 495–511. L. A. Kirkpatrick and P. R. Shaver, "Attachment theory and religion: Childhood attachments, religious beliefs, and conversion," *Journal for the Scientific Study of Religion* 29 (1990): 315–334.
23. J. Adler, "In search of the spiritual," *Newsweek*, August 29, 2005, 46–64.
24. W. L. Gardner, C. L. Pickett, V. Jefferis, and M. Knowles, "On the outside looking in: Loneliness and social monitoring," *Personality and Social Psychology Bulletin* 31, no. 11 (2005): 1549–60.
25. R. Kraut, M. Patterson, V. Lundmark, and S. Kiesler, "Internet paradox: A social technology that reduces social involvement and psychological well-being?" *American Psychologist* 53 (1999): 1017–31.
26. L. H. Powell, L. Shahabi, and C. E. Thoresen, "Religion and spirituality: Linkages to physical health," *American Psychologist* 58 (2003): 36–52.
27. A. Lamott, *Bird by bird: Some instructions on writing and life* (New York: Anchor, 1994).
28. Mitchell quoted in K. Kelley, *The home planet* (Reading, MA.: Addison-Wesley, 1988), 138.
29. Center for Budget and Policy Priorities, quoted in E. Konigsberg, "A new class war: The haves vs. the have mores," *New York Times*, November 19, 2006.
30. D. Brooks, "Mourning Mother Russia," *New York Times*, April 28, 2005.

31. K. Vohs, N. Mead, and M. Goode, "The psychological consequences of money," *Science* 314, no. 5802 (2006): 1154.
32. G. A. Kaplan et al. "Inequality in income and mortality in the United States: Analysis of mortality and potential pathways," *British Medical Journal* 312, no. 7037 (1996): 999–1003. See also B. P. Kennedy, I. Kawachi, and D. Prothrow-Stith "Income distribution and mortality: Cross-sectional ecological study of the Robin Hood index in the United States," *British Medical Journal* 312, no. 7037 (1996): 1004–07.
33. L. F. Berkman and I. Kawachi, *Social epidemiology* (Oxford: Oxford University Press, 2000), 164.
34. R. Axelrod, *The evolution of cooperation* (New York: Perseus, 2006).
35. J. Donne, *Devotions upon emergent occasions* (1624), in *John Donne: The Major Works* (New York: Oxford University Press, 2000), 333–351, 344. C. Darwin, *The descent of man and selection in relation to sex* (1874; Chicago: Rand McNally, 1974), 119.
36. M. A. Nowak, "Five rules for the evolution of cooperation," *Science* 314 (2006): 1560–63.
37. C. S. Lewis, *The Four Loves* (New York: Harcourt, 1960), 2. E. O. Wilson, *Consilience* (New York: Vintage, 1999), 6.

index

Page numbers in *italics* refer to illustrations. Page numbers beginning with 271 refer to endnotes.